R. Rawlins

A Dissertation on the Structure of the Obstetric Forceps

pointing out its defects, and especially of those with double curved blades - at the same time, shewing particularly the safe application of those with single curved blades

R. Rawlins

A Dissertation on the Structure of the Obstetric Forceps
pointing out its defects, and especially of those with double curved blades - at the same time, shewing particularly the safe application of those with single curved blades

ISBN/EAN: 9783337780777

Printed in Europe, USA, Canada, Australia, Japan

Cover: Foto ©berggeist007 / pixelio.de

More available books at **www.hansebooks.com**

A
DISSERTATION
ON THE
OBSTETRIC FORCEPS.

A
DISSERTATION
ON THE
STRUCTURE
OF THE
OBSTETRIC FORCEPS,
POINTING OUT ITS DEFECTS,

AND ESPECIALLY OF THOSE WITH

DOUBLE CURVED BLADES:

AT THE SAME TIME,

SHEWING PARTICULARLY THE SAFE APPLICATION

OF THOSE WITH

SINGLE CURVED BLADES,

AS GEOMETRICALLY

PROPORTIONED AND CONSTRUCTED:

AND LIKEWISE,

Shewing the Necessity and good Effects of several new Forms of the SINGLE CURVED BLADE,

AS THE

NARROW, FANGED AND REFLECTED,

IN

CERTAIN CASES OF RETARDED LABORS:

TOGETHER WITH

Cautions, Remarks, and Reflections on the Conduct and Management of Labors in general.

BY R. RAWLINS, SURGEON, Oxford.

Festina lentè———
Sat cito, si sat bene———

LONDON:
PRINTED FOR THE AUTHOR;
AND SOLD BY B. AND J. WHITE, FLEET-STREET.

1793.

TO

MARTIN WALL, M.D.F.R.S.

FELLOW OF THE COLLEGE OF PHYSICIANS, LONDON;
AND
LITCHFIELD'S CLINICAL PROFESSOR OF PHYSIC
IN THE UNIVERSITY OF OXFORD.

As a testimony of regard, with the sincerest thanks for his attention and great benefit received in several cases of ill health from his medical knowledge and judgment; and likewise, for the great respect to his qualifications and abilities, in Literature, Science, Arts, &c. in general, this Dissertation on the Structure, &c. of the Forceps, as adapted for the Delivery of Women in certain Cases of retarded

tarded Labors, is very gratefully, and with all due regard, respect, &c. humbly inscribed,

By his most obliged Servant,

THE AUTHOR.

OXFORD,
March 1, 1793.

To the PUBLIC.

As an apology for this publication the Author thinks it neceſſary to obſerve, that whoſoever ſuppoſes he has made an improvement in any art, &c. ought undoubtedly, as a good citizen, to communicate it to the world for the benefit of his fellow creatures.

Actuated by this principle, and induced by thoſe of humanity, to ſave the lives of infants; to mitigate and alleviate the pain and diſtreſs of women in parturition; to prevent their receiving any injury during that ſtate, and conſequently to promote their happy, ſafe, and perfect recovery; and alſo to obviate anxiety in the mind of every good, careful, and humane practitioner:

tioner: the Author publishes this Dissertation, regardless of censure as to its incorrectness or inelegancy of language. It is truth only he wants to express, and that in the most natural and plain manner, though perhaps with some necessary tautology.

ERRATA.

Page 4. Line 19. *for* that it evidently shews, *read* as evidently to shew.
——— 10. ——— 1. *for* ends, *read* end.
——— 26. ——— 1. *for* may, *read* must.
——— 31. ——— last, *for* nay, *read* though.
——— 44. ——— 9. *for* tinctura, *read* infusum.
——— 53. ——— last, *for* thus happily the hæmorrhage ceases, *read* then happily the hæmorrhage will cease.
——— 64. ——— 8. *for* will with more ease, *read* will gradually with very little pain.
——— 66. ——— last, *for* as prolapsus uteri, *read* as procidentia vel prolapsus uteri.
——— 109. ——— 13. ⎫
——— 110. ——— 6. ⎬ Draw a small stroke with your pen under the letter a: thus a
——— 111. ——— 8. ⎭
——— 112. ——— 6. *for* be so regularly or to so good a purpose accomplished, *read* so regularly or to so good a purpose be accomplished.
——— 116. ——— 21. after the word length, *add* reflection.

A

DISSERTATION

ON THE

OBSTETRIC FORCEPS, &c.

───────

TO

DR. WALL.

SIR,

PERMIT me to lay before you a short differtation on the ftructure and application of an inftrument, called the Forceps, as adapted for the delivery of women, in certain cafes of retarded* labors; with fome cautions, remarks and reflections, as to the conduct and management of labors in general.

* Thofe commonly called tedious, lingering, laborious, difficult, &c. I call here (and fhall hereafter clafs as) retarded labors, as evidently depending on fome caufe retarding the natural expulfion of the child.

B With

With respect to the real inventor of the Obstetric Forceps, I will not pretend to say who he was, but I am confident that he deserves the highest praise, and the sincerest and most grateful thanks from every woman; for, undoubtedly, he very humanely considered the painful, and often deplorable state of women in labour; and shuddered at the indecent modes that were then used, and at the cruel, horrid, and often ineffectual means, that were at that time taken to deliver women, and to extract the child.

As a good, careful and attentive practitioner, he not only wished to shorten and mitigate the agonizing pains of labour, but he also very wisely considered the great importance of saving a child's life; to effect these purposes he must have very attentively compared the shape of an infant's head with the figure of the female pelvis, and as judiciously supposed, that, if in certain cases of retarded labors the child's head could be inclosed or encompassed between two pieces or blades of smooth polished iron semielliptically bent, the longitudinal shape of the child's head might gradually be drawn down,

down, and as effectually be turned into the long or tranfverfe diameters of the female pelvis, and thus the child might be fafely delivered alive, and that without any injury to the mother.

Such then, it is natural to fuppofe, were the firft fuggeftions or thoughts of the real inventor of the Obftetric Forceps,; but, let this be as it may, it is evident, that the invention of the Forceps was much earlier than it is in general imagined, though commonly afcribed to Dr. *Chamberlain*; but of this there is no pofitive proof, and indeed it is fomewhat evident, that he was not the real inventor of it; for though *Chamberlain* and his fons followed the practice of midwifery, they only afferted, like many other practitioners in thofe days, that they were in poffeffion of a fecret inftrument, by which they could readily deliver, beyond all other practitioners at that time, with the utmoft fafety to the mother and child, whatever lingering, laborious, or dangerous cafe of labor the woman might then be fuffering under; but notwithftanding all this boafting, in a particular cafe of this kind, one of the *Chamberlains* was foiled at Paris, whither he went to perform prodigious feats of delivery beyond, in his own imagination, what the *French accoucheurs* could do. Now this very failure of *Chamberlain*'s son evidently proves, even though his inftrument had been the Forceps, that he knew nothing of the principles of it, and it likewife proves that his family was not the real* inventor of it; for if it had been

---

* It may here be acknowledged, that the inventor of any kind of inftrument may not know all the real ufes and applications of it, he may only have invented it under his own particular idea to anfwer that one purpofe; fo it may be faid that *Chamberlain* may have only invented the Forceps merely under the idea of laying hold of the child's head, and thus to extract it however forcibly; but I cannot think fo, for the Forceps in its ftructure is fuch, ~~that it~~ evidently, fhew, that the real inventor of it muft have had more ideas than this, for he muft have in its conftruction very minutely confidered the fhape and fize of the child's head, and ftructure of the female pelvis, and accordingly have applied them together, and fitted the figure of the Forceps thereunto, confequently muft have known the true application and ufe of it, whereas the *Chamberlains* did not,

for

been so, the *Chamberlain*, who went to Paris, would have been properly acquainted with the figure or shape of the child's head, and have known perfectly well the structure of the female pelvis, consequently would have compared them together, and would have then known, in the case he attempted to deliver at Paris, the true presentation of the child's head, and the circumstances of the mother, whether the pelvis was distorted or not, and accordingly would have omitted, or would have properly applied his instrument so as not to have been foiled.

But without meaning any censure on the *Chamberlains*, I shall declare, that he who attempts to keep any important invention, as the Obstetric Forceps certainly is, a secret from the world even during any part of his life, deserves not the honour of it; but he only who generously publishes it for the general good of his fellow creatures:

for by their practice it seems that they introduced their instrument at random, and extracted the child's head by mere force.

hence the merit of the invention of the Forceps for the delivery of women in certain cafes of retarded labors, ought to be given to Mr. *Chapman*; for it was he who voluntarily firſt gave it to the world, though after deſcribing its form and explaining its uſe, he with great modeſty declares, that it was the ſecret inſtrument of the *Chamberlains*, but yet this does not prove *Chamberlain* to be the real inventor of the Forceps; however admitting it, though there is much to be ſaid againſt it, it is nevertheleſs evident, that the Forceps, as deſcribed by *Chapman*, had many defects; for, firſt, its blades were too broad, which hindered them from being eaſily and readily introduced. And ſecondly, their curvature was too circular, which, on any very little preſſure of the handles, would occaſion their ends to lacerate or bruiſe the ſides of the infant's head. And thirdly, beſides theſe defects, the two ſides of the Forceps were to be joined or faſtened together by means of a ſcrew, and which could not always be accompliſhed; ſo that from this laſt

laſt defect, it is but natural to ſuppoſe, that in ſuch caſes requiring the uſe of the Forceps, practitioners at that time, though very injudiciouſly ſometimes, inſtead of ſcrewing the ſides of the Forceps together, tied the two ſides together with a ſtring, &c. from which many evils and much injury to the woman muſt have frequently ariſen, either from the unſteadineſs of the blades, their turning aſide, or ſlipping off, and thus materially hurting the internal parts of the vagina, &c. But fourthly, the greateſt defect of *Chapman's Forceps* was in its length, not only of its blades, but more eſpecially of its handles, which undoubtedly were too long; for altogether it gave them too much power, not in extracting the child's head, but in their lever, by which they were very liable to produce a violent compreſſion at their fulcrum, on the bones of the pelvis, when the handles were ever ſo little inclined out of the deſcending axis of the pelvis; hence much injury to the woman.

With all theſe defects, the Forceps with

very little alteration or amendment continued until Dr. *Smellie*'s time, for it was he who very ingeniously and prudently contrived to lock the sides of the Forceps together in a very handy, convenient and steady manner; he also lessened the curvature of the blades, and very judiciously shortened not only the blades, but also very materially the handles; for it must be observed, that in delivering with the Forceps, the handles should be so used, as never to occasion any pressure from the blades on the bones of the pelvis, for whenever that happens, as from the forcible and injudicious working of the handles of the Forceps from * side to side in the extracting of the child's head, a certain point on the vagina, and other parts contiguous and lying beneath it, will be exceedingly compressed, bruised, or lacerated, by such absurd and violent actions of the Forceps; and hence from such contusions, &c. much mischief will arise, as inflammation, ulceration, &c.

* As generally directed by most authors, and followed by most practitioners. *Horrid actions!*

nay mortification, and even death itself; all which short handles, &c. to the Forceps will generally, if not always, prevent.

Hence, to avoid such accidents and misfortunes, Dr. *Smellie* thought it proper to alter the shape, length, curvature, &c. of the Forceps then in general use; but, nevertheless, though *Smellie* very carefully and judiciously altered the Forceps in almost every part, yet, as he gave not curve enough to the blades (the distance between them in their middle, when the two sides were locked together, being little more than two inches), their ends in consequence thereof came nearly close together, so that, therefrom, they are not only very apt to lacerate, but may compress the head so violently as even to kill the child, particularly in the hands of the hasty and injudicious practitioner, who very inattentively too closely and strongly squeezes the handles together in extracting the child's head.

But these accidents may in a great measure be prevented, and these defects of *Smellie's Forceps* may be removed, by giving the

the end of each blade an expanfion of the four tenths of an inch from the axis of their handles, and alfo by curving each blade nearly in their middle to the diftance of an inch and the four tenths of an inch from the faid axis, and that merely by a little more fuddenly bending outwardly the fhank of each blade juft above the groove of the locking part, and then immediately therefrom making the curve of the blade nearly to correfpond to the fegment of a quadrant, whofe radius is about four inches and a half; fo that when the two fides of the Forceps are locked together, the blades fhall ftand feparate from each other at their ends about the diftance of eight tenths of an inch, and nearly in the middle of their curvature about two inches and eight tenths of an inch, that is, nearly three inches; for clofer than this no child's head of a common moderate fize can with fafety be compreffed, and if the ends of the Forceps' blades are fet wider, and their curvature lefs, the Forceps would ever be liable to flip off. Hence whenever in future I fhall mention the fingle curved Forceps, I hope to to be understood that I mean Dr. *Smellie*'s common Forceps thus altered, and which I shall at the latter end of this Dissertation attempt geometrically to construct and delineate as in Fig. 1 and 2.

But further with respect to the Forceps it must be observed, that since Dr. *Smellie*'s time, Dr. *Osborn* has also thought it proper to make some other alterations in the blades of the Forceps, not only in extending their length near half an inch, but also in expanding them to the distance of three inches in the middle of their curvature, and setting their ends to the distance of the six tenths of an inch when the sides are locked together; and besides this, he has also very considerably widened the blades in their breadth to one inch and the six tenths of an inch, and gave them a kind of double, or rather a triple curvature (in imitation of *Leveret's Forceps*), gradually inclining from the shank just above the locking part, to almost the top of each blade, nearly to the distance of two inches from the common axis or straight line through the middle of the handles; and to make these alterations

tions it seems that he has been induced, from a supposition that the blades thus formed are more convenient to apply*,
more

---

* How far Dr. *Osborn* has found the double curved Forceps preferable to the single curved Forceps I cannot say, but I think that he is much mistaken therein; the reason which he teaches for such double curved blades being more conveniently applied, is that of not requiring the perinæum in their introduction to be so far distended backwards as the single curved do; but this is not absolutely true, for in those presentations requiring one blade to be introduced towards the pubes, and the other towards the sacrum, the perinæum must then, for their proper introduction and safe application, be drawn equally as far backwards as if the single curved blades had been introduced; for the double curve, being in a contrary direction with that of the handles, cannot have the effect which the Doctor mentions; and besides, it must be observed, that when a woman has been in labour so long, as the case to require the assistance of the Forceps, the perinæum is then generally so relaxed as to distend backwards easily, even further than what is generally wanted to introduce the single curved Forceps, hence this reason is vague and futile.—And further, it is very evident that the blades, when double curved, cannot in every presentation more aptly fit the head than the single curved; in some presentations of the face they may perhaps be equally so, but in others the single curved blades are in this very respect by far preferable. And, with respect to the double curved blades being more safe
and

more aptly fitting the child's head, and consequently more effectual in extracting it.

But and effectual in the extraction of the head than the single curved blades, I should think, on a very little reflection concerning the axis of the double curved Forceps, that it would immediately appear to every one to be a mistake; for, let us suppose that in extracting the child's head with the double curved Forceps, the chief hold of the Forceps must certainly be very near the ends of the blades, consequently the axis on which the power in extracting depends, must proceed in a straight line across the curvature from the ends of the blades to the handles, which is in fact the direct action of the single curved Forceps.—What advantages or superiority then has the Forceps with double curved blades over that which is single curved in its blades? Certainly none; but, on the contrary, the double curved Forceps has its defects, as it may, for the reason just given, very much deceive the operator in the axis in which he is to direct the power of his instrument when extracting the head, and therefore, instead of facilitating the delivery, may retard it, and render it more painful, as drawing and pressing the head into a contrary direction to what the operator intended; and again, it may deceive the operator even in introducing the blades; for as it is the invariable rule with Dr. *Osborn* always to place the convex edge of his Forceps towards the infant's face, so in face presentations towards the pubes, in thus introducing such double curved blades, their ends will be very apt to be forced,

unthinkingly,

But notwithstanding all these alterations in the Obstetric Forceps, and although granting that the double curved Forceps of Dr.

unthinkingly, against the upper part of the sacrum or projection of the lumbar vertebræ, and there hurt the woman, and this merely owing to the double curvature of the blades deceiving the operator in the axis of the Forceps: whereas the single curved Forceps would freely pass up over the ears of the child in the middle or direct axis of the pelvis, consequently not so liable to injure any part therein. But not to mention at present any more such instances which tend to prove that the double curved Forceps is not by far so safe and effectual as the single curved Forceps in many cases of retarded labors, I shall now only point out an omission of Dr. *Osborn* respecting his double curved Forceps, and then any candid practitioner may readily judge how very attentively the Doctor has considered the structure and application of his double, or rather triple curved Forceps; in the engraving of it in his plate, he has only given us the view of one side of it, not considering that, if any future or foreign artist was to attempt to make a pair of double curved Forceps therefrom, the ends of the two blades, when locked together, would point different ways, instead of being opposite to each other; this Dr. *Osborn* will find to be true, if he will be so kind as once more to take a view of the different curvatures of his Forceps' blades, one evidently inclining over the groove of the locking part, the other turning from it; it therefore behoves the Doctor, in his next or second edition of his

Dr. *Osborn*, and the single curved Forceps of Dr. *Smellie* altered as just now mentioned and described, page 10, and shewn in Fig. 1 and 2, may be very applicable, safe and effectual, in many cases of retarded labors, yet there are certain other cases of this kind, in which their blades cannot be very readily introduced, or very safely applied; therefore further to remove these defects in the Obstetric Forceps, and to render the blades more convenient and easier to be introduced, I have thought it proper to contract the breadth of one blade of the single curved Forceps into a narrow blade not more towards its end than one inch, as in Fig. 3. And I have also thought it proper to divide one blade of the single curved Forceps into two separate parts, the fangs of which are each in breadth about three

his Essays on the Practice of Midwifery, to give a better description of his Forceps than he has, and likewise an engraving of the two sides, and particularly of the blades, instead of referring his readers to Mr. *Savigny* or Mr. *Carsberg*, instrument-makers in London, for a sight of his Forceps, as made by them exactly, as the Doctor says, to his own directions, or rather to their directions.

tenths

tenths of an inch, as in Fig. 4 and 10, and at the point of the blade's bifurcation have joined them together with a hinge; by the contrivance of which, and by the handles of the fangs being bent and made to diverge off at the hinge, the two fangs may be kept so close together, by the finger being placed between their handles, as to be in effect but one narrow blade in breadth the six tenths of an inch, and thus it can be easier, more readily, and conveniently introduced under the arch of the pubes than the blade either of the single or double curved Forceps, as taking up less room, consequently will more readily pass up the space between the ossa pubes; and further, when the fangs are introduced by removing the finger that was placed between their handles, and then on pressing their handles close together, the two fangs will be immediately spread open, and cannot close again, when the two sides of the Forceps are locked together; for the projecting cheek of the groove for the locking part of the opposite blade keeps the double handle of the fangs perfectly secure, safe and firm, so that the two fangs will remain steadily fixed* and expanded.

Hence with these kinds of Forceps, viz. the single curved, the narrow, or fanged blade, as adapted severally to the particular cases, the delivery of almost every variety of retarded labor may be very safely and effectually accomplished, without injury to the mother, or hurt or destruction to the child.

But to be a little more explicit with respect to the use of the above-mentioned kinds of Forceps.—Let it then be observed, that though Dr. *Osborn* supposes the double

---

* I am sensible that some practitioners will say, by the division of one blade of the single curved Forceps into fangs, that in introducing them within the projection of the pubes their ends (or as they may be termed points) will be more apt to wound, pinch, nay even pierce through the vagina, &c. or lacerate the child's head, than the more blunted ends of either the single or double curved Forceps—granted. But such will be the case with every kind of Forceps in the hands of the rash and impetuous practitioner, who boldly attempts with force to introduce the blade against any obstruction to it, instead of letting it pass on freely, easily, readily, and without force, and immediately stopping and withdrawing it on meeting with any resistance.

curved Forceps to be more effectual and safe than the single curved Forceps in all cases of retarded labors, as particularly when the face presents itself forward with the chin towards the pubes, with the vertex of the head descending into the hollow of the sacrum, or even when it has got no further than just under the projection formed by the upper part of the sacrum and curvature of the lumbar vertebræ; and also, when the face presents itself forwards, with one ear to the pubes and the other to the sacrum, and resting as it were on the spinous processes of the ischia, and cannot from its position proceed any further; or when the vertex of the head presents with one ear to the sacrum and the other to the pubes, and although the pelvis is duly formed, of its proper size and elliptical figure, yet from the largeness or length of the child's head it rests on the spinous processes of the ischia, or perhaps higher up in the pelvis, and can proceed no further, nevertheless, in such like cases, I cannot but think that every practitioner, after impartially considering the structure of

Dr.

Dr. *Osborn*'s double, or rather triple curved Forceps, and comparing it attentively with the shape of the child's head and mechanism of the female pelvis, and carefully examining the different circumstances attendant on such labors, will very candidly confess, that the double curved Forceps is not more safe, preferable and effectual, than the single curved Forceps; nay, I rather believe, that they will decidedly determine that the double curved Forceps is not so effectual, safe, nor so well adapted for the delivery of women in such cases of retarded labor as the single curved Forceps is; but yet it must be confessed, that in the hands of the expert and judicious practitioner, perhaps any kind of Forceps may be very safe in effecting the delivery; but, in the hands of the inexpert, unthinking and hasty practitioner, the double curved Forceps will certainly do much mischief.

But, however, it sometimes happens, even in the most natural presentations of a common sized head, that one blade of the single curved Forceps, or the blade of any other Forceps heretofore invented, cannot be easily or readily introduced between the head

head * and pubes, nay not without some confiderable force, and fometimes cannot by any means be introduced, and this in fome meafure owing to a too great ftraitnefs of the facrum obftructing the child's head in its natural defcent through the pelvis, but chiefly to a too forward projection of the pubes diftorting the natural figure of the female pelvis, rendering it rather of a triangular figure, as in the male pelvis. Hence, by the too forward projection of the pubes, the fpace between them is rendered too narrow to admit the breadth of the blade even of the fingle curved Forceps, fo that when the blade is introduced at the os externum, the end of it from its too great breadth cannot pafs up between the head and pubes, but will go directly againft the head, and hence from this obftruction of the head in fuch cafes to the introduction of the Forceps' blade, the

---

* In fuch cafes fome practitioners recommend introducing one blade of the Forceps firft towards the facrum, and then to bring it round to the pubes; this is certainly dreadful practice, and more likely to injure the uterine parts than the careful and gradual introduction either of the blade divided into fangs as Fig. 4 and 10, or the narrow blade Fig. 3.

head has often been deemed too large to pafs any further, and confequently has been, by the too hafty and rafh practitioner, unthinkingly and very precipitately opened, and the child deftroyed, in order to facilitate the delivery of the woman.—Now all this may be avoided by the introduction of the contracted narrow blade Fig. 3, or the fangs of the divided blade Fig. 4 and 10 of the Forceps, as juft mentioned, which, from their narrownefs, can in fuch cafes fall further in between the too clofe and forward projection of the pubes, and confequently may more eafily, readily, and without hurting the woman or child, be introduced; and when introduced, the two fangs of the divided blade will immediately open, and when locked will become in action equal to the blade of any other Forceps; for it muft be obferved, that in extracting the head in the above cafe with the Forceps, the chief power and action is from that blade next the facrum, that next the pubes ferving only as a ftay or fupport and director to the other; for as the head is brought, or rather drawn down, and descends in the pelvis, the action of the hindermost blade, or that next the sacrum, forces and turns the vertex of the head towards the arch of the pubes, so that the face falling into the hollow of the sacrum, the long shape of the child's head is turned into the lower and long part of the pelvis, from whence the head may be easily and readily delivered; hence, by rendering one blade of the single curved Forceps so narrow as to be readily introduced between the pubes and the child's head in such retarded labors, occasioned in some measure by too great a straitness of the sacrum, as just mentioned, but chiefly by a too forward projection of the pubes, distorting the due size and natural elliptical form of the female pelvis, particularly towards the pubes, many advantages must arise, much trouble, difficulty and anxiety to the practitioner may be avoided, and, what is more particularly to be considered, the destruction* of the child may be often prevented,

---

* These advantages certainly will set aside every fear as to the use of the narrow or fanged blade of the forceps wounding, pinching, or piercing through any internal part

vented, and that from the use and application of the single curved Forceps with one blade divided into two fangs, as Fig. 4 and 10, or contracted into a single narrow blade, as Fig. 3.

Such then, I think, ought to be the necessary forms or structure of the different kinds of Forceps, that should be applied in the different cases of retarded labors for the safe and happy deliverance of the child without injury to the mother.

But further with respect to the use of the Forceps, it is necessary to observe, that in whatever kind of retarded labor the use of any kind of Forceps is required, particular care must be taken that, in extracting the child's head, the action of the handles of the Forceps be not directed out of the descending axis of the pelvis; for when that happens, it throws the natural diameter or shape of the child's head out of

part of the woman, and particularly so, as every contrived method for delivery should be first attempted before the practitioner should dare to open the head, and thus to destroy the child.

that descending line*, and thus not only retards the labor, and renders the head more difficult to be extracted, but the blades of the

* As by the injudicious and forcible working or motion of the handles from side to side in the extraction of the child's head, so generally recommended by authors, and so often practised in such operations; for it must be observed, that whenever the handles of the Forceps in extracting the head are so used, that then each side of the Forceps acts alternately as a vectis or lever, and therefore the Forceps must then, in fact, be considered as a double vectis, and that whenever its handles are moved from side to side, each blade must become a vectis alternately, whose fulcrum, as relative to the woman, is not at the joint or locking part as is generally imagined, whatever it may be as to the child's head, but on the convex side of the blade which on ever so little inclination of the handles of the Forceps out of the descending line or axis of the pelvis will immediately press on the bones of the pelvis, and if the handles are moved forcibly out of their proper line, the fulcrum, which then falls on the convex side of the blade, to which side the handles are inclined, will greatly compress and contuse the soft parts that may then lie between the blade and bones of the pelvis; hence much mischief frequently arises from such motions or working of the handles from side to side in the extraction of the child's head.—That the fulcrum of the Forceps is not at its joint or locking part is extremely evident, as there is no fixing the opposite the Forceps immediately compressing on the sides of the pelvis, whatever part is between them and the bones of the pelvis will be greatly contused; hence much mischief will certainly ensue; therefore such undue inclinations of the handles of the Forceps must be carefully avoided, and hence for the safe and easy extraction of the head, in every action of the Forceps, the descending line or axis through the pelvis must be particularly observed, so that the handles of the Forceps be not carried out of it, by which means the long shape of the head will be gradually and regularly drawn down, and turned into the long or transverse diameter of the inferior or lower part of the pelvis. And further,

posite blade for it to become such; for though the blades are each strongly squeezed and firmly attached to the head by the operator firmly grasping and pressing the handles, yet as the head is moveable from side to side, the fulcrum cannot fall on the joint as generally supposed, but alternately on the convex side of the blades as they are moved from side to side, the moment they press on the bones of the pelvis. How cautious then ought every practitioner to be in extracting the head, that he inclines not the handles of the Forceps out of the descending axis of the pelvis!

it ~~may~~ *must* be observed, that when the face or vertex of the head is brought so low as to point directly through the os externum, and the other parts of the head rest on the perinæum, that then the blades of the Forceps should be taken off* and removed, and the head left entirely to be expelled by the natural efforts of the woman's pains; by these means the perinæum will be naturally distended, and a laceration of it

---

* In general it is directed, that the child's head should be entirely extracted with the Forceps, in very weakly women where the pains are too slight and insufficient to protrude the child, and where also the perinæum is in a lax, yielding state, the labor may be so far finished, and the head extracted slowly, gradually, and regularly; but when the perinæum is rather rigid and not very yielding, and the woman of a healthy robust constitution, and the pains strong, it is much better to remove the Forceps as soon as the head begins to dilate the os externum, than to extract the head forcibly with the Forceps, for fear the perinæum should be lacerated; and it will be more agreeable both to the woman and to the bystanding women, when they perceive that you act with caution and deliberation, and not with force; and besides, it is always then in our power to refix the Forceps instantaneously, and that with ease and facility, if any unforeseen accident should happen to arise, which I believe never will.

(which

(which too commonly happens if the head is forcibly drawn out by the Forceps) will be prevented; for such disagreeable accidents to women are entirely owing to force, or to an insufficiency of time being allowed for the texture of the perinæum and parts round it to relax, or to dispose themselves for a proper and necessary dilatation; therefore, if a proper time is not given, after the vertex or face is brought so low as to point directly through the os externum, for the fibrous texture of the perinæum, &c. to dilate and extend itself backwards, a laceration of it will certainly be the consequence, if the head is forcibly and precipitately drawn through the os externum with the Forceps, as is often done by the hasty and impetuous practitioner.

As much mischief then may be done by too hasty, bold, and imprudent practitioners, it will be necessary to remark, that in the use of the Forceps, the blades must be introduced slowly, readily and gradually, without force, that they be easily and evenly locked together, so that the opposite blades may exactly correspond with each

each other; and when locked, the handles should be only so compressed together, as for the blades to take such a moderate firm hold of the head as just sufficient to keep them from slipping off; for if they are closely and strongly compressed together, the head will be lacerated, or so violently bruised and pressed between the blades, as perhaps even to kill the child. And further to remark, that in extracting the head, it should be performed with as little force as possible, not even if possible with more than merely holding down, or retaining the head in the position in which the preceding labor pains of the woman had placed * it; observing

---

* Though some practitioners pretend to say that force is absolutely necessary, and sometimes ought to be so far exerted as even to lift the woman up from her bed with the Forceps when fixed on the child's head within her; *O horrible work!*—But yet let it be observed, that such force is not necessary, for as the bones of an infant's head are not perfectly nor entirely ossified nor connected together but by membranous substances forming the sutures and fontinells of the head, so the head of an infant is thereby, through compressure, rendered capable of being changed in shape and diminished in size, and thus may be so elongated and adapted to the cavity of the pelvis,

serving at the same time, as the head descends in the pelvis, that, in making the turn necessary for placing the long shape of the child's head into the long transverse diameter of the lower part of the pelvis, the direction of the Forceps' handles be equally, slowly, gradually, and regularly inclined, and steadily kept in the descending axis of the parts, and that with as little force as possible, by which means every kind of injury to the mother will be avoided, and the head very readily and safely brought down so low as to have its crown rest on the perinæum, and its vertex pointing out of the os externum; and now, as the head has passed by every resisting cause, there can be no further use for the Forceps; therefore it must be ever remembered, that, when the vertex of the head or face is

vis, as to take its shape, and to pass through it with very little force, provided the practitioner will give for it a proper time, and is not harsh or impetuous in extracting the head with the Forceps. In short, whatever force is necessary, it should be begun in the most slight and gentle manner, and not wantonly increased, or even exerted beyond what will gradually, regularly and almost imperceptibly, extract the head.

brought

brought so low as to dilate and point out at the os externum, and the crown of the head to distend and rest on the perinæum, the blades of the Forceps should be taken off, and the remaining part of the labor, as before mentioned, should be left entirely to nature, in order to prevent a laceration of the perinæum, a misfortune very disagreeable to women, and therefore ought, by every means, attention and care, to be prevented, if possible.

Thus far I have thought it proper to speak of the different forms of the Obstetric Forceps, with cautions, remarks, &c. necessary to be observed as to the use and management of their particular application in certain cases of retarded labors.

I AM next to proceed and make some reflections, remarks, &c. and give some cautions respecting the conduct and management of other kinds of labors, in which the assistance of the Forceps is not so materially nor so especially necessary.

And first,

Of those kinds of retarded labors in which the head cannot be extracted by the Forceps, but in order to save the woman's life, it is requisite to come to the last resource, that of opening the skull of the child's head, so that the head may be reduced in size sufficient, with the assistance of the crotchet, or blunted hook, &c. to pass through the pelvis, let whatever may be its figure, shape, or distortion.

It will therefore be necessary to say, in every case requiring the opening of the child's head, to reduce it to a size proper to pass through the pelvis, and to be extracted with the crotchet or blunted hook, &c. that there is no operation in midwifery that requires so much care or greater deliberation; for though it may be said, that such cases always proceed from the same cause, namely, that of the head being too large to pass through the pelvis, nevertheless the head may be itself so large and hydrocephalous, as even not to pass through a well-formed pelvis; and, on the contrary, *though* ~~nay~~

though the head may be of a natural fize, nay not fo large as common, yet the pelvis may be fo fmall and diftorted as not to admit it to pafs through. In the firft cafe, merely by opening the head and evacuating its contents, the head may be fo reduced in fize as foon after to defcend fo far down in the pelvis as to be readily extracted with the blunted hook, or even with the Forceps, or perhaps even forced through the pelvis by the natural labor pains. But in a diftorted pelvis the cafe is otherwife; there is more danger and difficulty, and that in proportion to its diftortion: wherefore it is neceffary to remark, that if the pelvis is not much diftorted after opening the head, it may be perhaps fo reduced, as after a little time to be protruded fo low down in the pelvis, as to be readily extracted with the blunted hook or crotchet*, &c. But if the pelvis

---

* To extract the head of a child after its fkull has been opened and its contents evacuated, fo as to reduce it in fome degree to a fize capable of being drawn through the pelvis, a variety of modes have been propofed, and inftruments invented and recommended, but
thofe

pelvis is so much distorted, that there is not two, or scarcely two inches and a quarter of an inch distance from the sacrum to the

those called the blunted hook and crotchet, have been hitherto for this purpose the most esteemed; nevertheless, even in the use of these, practitioners differ much in their opinions; but as it would be very absurd and tedious in this place to endeavour to refute any particular opinion concerning them, I shall only say, respecting their use in such cases, what appears to be fact. If then the pelvis is not very narrow, and the head descended somewhat low in the pelvis, the blunted hook is preferable, and its end, if possible, should be fixed over the jaw-bone, as affording the firmest hold, and not being apt to tear out; consequently no injury will happen to the woman; but then the operator must constantly observe, in extracting the head therewith, to use as little force as possible, and to keep the handle in the descending axis of the pelvis, instead of working it to and fro, and thus avoid hurting the internal parts of the vagina, &c. But if the pelvis is narrow, and the head not far advanced into it, the crotchet may then be preferable, and should always be introduced through the opening made in the skull, and its end carried into the foramen magnum, and there firmly fixed; and then taking every precaution in extracting the head, that the crotchet does not tear out and hurt the woman; but if the end of the crotchet cannot be readily fixed in the foramen magnum, it should not be fixed any where else, for probably it will tear out and hurt the woman, unless you guard her

the pubes, much difficulty immediately arifes therefrom; fo that after opening the head, its extraction muft be deferred for fome

very carefully with your other hand, and fo perhaps receive the injury yourfelf.

But however, to avoid in a great meafure any accident from the ufe of the blunted hook or crotchet, and indeed in feveral cafes to fuperfede the ufe of any fuch fort of hooked inftruments for extracting the head after it is opened, it has fuggefted to me to reflect backwards one blade of the fingle curved Forceps, as in Fig. 11, 12, and 13, the outward blade of which is always to be the narrow blade of the Forceps Fig. 3, which, as ufual, is to be introduced into the pelvis over the outfide of the head, but the reflected blade is to be introduced within the fkull, through the opening in it, and then on locking the two blades together, they will fo firmly include and hold the integuments and bones of the fkull as to enable the operator to extract the head equally as efficacious as with the blunted hook or crotchet; and therefore, as the reflected blade of the Forceps in its ftructure is fo much lefs liable materially to injure the woman than the end of the blunted hook or crotchet is, if peradventure either of them fhould very fuddenly and unexpectedly tear out, the reflected blade of the Forceps, as being thus preferable, ought, in every cafe of this kind, to be applied, if poffible. Hence in cafes requiring the head of the child to be opened for the delivery of the mother, after that operation has been performed, it will be proper firft to try to extract the head with the fingle curved Forceps Fig. 1,
and

some time, perhaps one or two days. or even longer, until some degree of putrefaction* comes on, when the bones of the head will readily separate from their integuments, and will perhaps so far protrude, that the greater part of them may be readily extracted with the fingers; and then to extract the remaining parts the blunted hook may be fixed over the jaw-bone, or the crotchet to the basis of the skull, after

and, if that is impracticable, then to try the reflected blade of the Forceps; and if that has not the effect, then to have recourse to the blunted hook or crotchet, always remembering not to introduce or fix either of these instruments in any uncertain manner, it being much better to leave the whole to nature and to the putrefaction of the head, and to get it away bone after bone, as the bones of the skull will slip from their integuments, and then the other parts of the child will soon follow.

* In a case of this kind I have known ten days to pass over, before the child, though in a very high degree of putrefaction, could be extracted, and yet the woman, though entirely senseless with cold clammy sweats, great difficulty of breathing, and pulse so weak as scarcely to be felt, for the six last days, contrary to every expectation recovered, so as to be at the end of her month as well as most women usually are; she has had no child since, as, on account of her dwarfishness, her husband and she very soon after by consent parted.

which the other remains of the child, as a soft yielding substance, will readily follow. Hence, great attention, care and caution, in such cases, are absolutely necessary, and positively required; and in short, let whatsoever be the figure or form of the pelvis, or size or shape of the head, it must ever be remembered, that after the operation of opening the head, and after a proper time waiting for its descending into the pelvis, that, in extracting it with the Forceps, blunted hook or crotchet, whether applied externally on the head or within the skull, it should be performed in a slow, regular and moderate manner, with as little force as possible; and that, when the child's body and shoulders are brought so low as to occasion a distention of the perinæum, &c. it should then be for some time left to nature, until a proper dilatation of the perinæum can take place; and when that will gradually yield and extend, then, if the natural labour pains are not sufficient to ~~exclude~~ *expel* the child, it may be slowly extracted, after which the separation of the placenta

placenta muſt be left to nature, and to be extracted as in a natural labor.

And now ſecondly,
It will be neceſſary to remark, that in conducting preternatural labors, the hand of the operator ſhould be introduced ſlowly, gradually, and with as little force as poſſible; and that when a foot or the feet of the child * are found and graſped with the operator's

---

* Whenever a child is to be extracted by the feet, ſome authors particularly adviſe, that the operator ſhould always with his hand ſeek for both feet of the child, and therewith extract it; but this is not always neceſſary, neither can it always be done without much danger, difficulty and pain, therefore much is to be left to the operator's judgment: if the pelvis is of a due ſize and form, it is not material whether the child is extracted by both feet or by one; ſo if both feet are not to be found readily, if the pelvis is large, one foot will be ſufficient, and the child may therewith be eaſily extracted: but if the pelvis is not very capacious, nay rather narrow, it may then be more prudent for the operator, if he can readily find the child's feet, to graſp them both and by them extract the child; ever remembering, that after having graſped a foot or the feet, if high up towards the fundus uteri, never to bring them backwards, but always to carry them forwards down to the belly of the child, by which

operator's hand, he should together with the child's foot or feet withdraw it in like manner; thus the body of the child will turn as gradually, and will be easily extracted; but remember, that, when the child is so far delivered as the buttocks to rest on the perinæum distending it outwards, then the operator should make a halt, and allow time for the texture of the perinæum &c. to yield, and so extend itself, that in the further extraction of the child's body * and head it may safely be delivered

which means the child will readily turn without injury to it, and may be perhaps delivered alive; but if the foot is imprudently reflected backwards, the child will not so easily turn, and perhaps either its thigh-bone will be broke or dislocated, or the vertebræ injured, and the woman much hurt.

* In extracting children by the feet, it is always necessary for the operator to observe, that, when the buttocks of the child are protruding through the os externum, in extracting the body, he should incline it so, that the belly of the child should turn towards the back part of the mother, by which means the breadth of the shoulders will turn into the long transverse diameter at the brim of the pelvis, and thus be easier brought down; and further, it is to be observed, by this inclination of the child's belly towards the back part of the mother,

that

delivered without hurting or lacerating the perinæum: and besides, thus preventing by this halt any laceration of the perinæum, we also obtain a proper time for the natural separation of the placenta as in a natural labor.

Again, thirdly,

It will be necessary to observe with respect to the delivery in flooding cases, that the management of them requires the greatest degree of attention, care, caution and deliberation.

Although, in the earlier months of uterogestation, if an uterine hæmorrhage should happen, nothing more can be done than to enjoin, in the strictest manner, rest, an horizontal supine posture of body, with slight restringents, cool air, &c. and after a few days, though sometimes not for a few weeks, as a blighted fruit, the abortive em-

---

that as the body and arms of the child are extracting, the face of the child turns also towards the back, and then the length of the head falls naturally into the same diameters of the pelvis, and thus turning with the face towards the sacrum, it is readily and easily extracted.

bryo, &c. will be discharged; and then, by the woman continuing in bed and at rest, with soft thin gruel, bread food, &c. for several days, until all appearances of discharge have stopt, all will be well.

But it is in the later months of utero-gestation that the greatest danger occurs, requiring the utmost circumspection how to manage them; if then, near to the time of parturition, a shew of blood, as it is called, should appear, the very first injunction is rest, and by keeping strictly a supine horizontal position of body, quietness of mind, with thin soft gruel, light bread diet, &c. with moderate restringents, cool or rather cold air, and a total forbearance from any attempts towards delivery, perhaps all in a little time may be well again. But if after the discharge has ceased for some hours or days, it should again appear, the same mode of treatment must be continued; for generally from the blood coagulating, and thus stopping up the orifices of the bleeding vessels, it will again cease, perhaps, for some hours or days, or perhaps remain very trifling for several days, after which

which it may again recur more copiously; and indeed, it cannot be altogether expected to stop, as it generally proceeds from a separation of some part of the placenta lying over the lower, though now the expanding part of the uterus; hence a separation of their vessels, and thence an hæmorrhage; therefore rest, slight restringents, cool or cold air, with soft food, &c. are as yet the only remedies; the practitioner indeed may touch his patient, but not with a view or thought of promoting or effecting a delivery, but to know the disposition of the os uteri; for nature, if possible, must be the only operator; the practitioner must be her servant and attend; he should, therefore, watch daily the dilatation of the os uteri, how far it has expanded itself, how thin grown, soft and yielding, and how great a portion of the placenta lies contiguous or over it; and he should carefully observe whether any slight pains at times occur, which shew that the fundus uteri is disposed to contraction, or has begun to contract; for without a constant attention being properly paid to these symptoms, the

practitioner

practitioner will find himself very soon at his *ne plus ultra*; for, though the discharge may be often very profuse, so that the woman may be as often faint, &c. nevertheless the practitioner must not be too hasty, he must have patience; the delivery must still be suspended, until some more sufficient pains come on, as they are of such consequence to the woman's life; for without some motion or contraction in the moving fibres of the uterus, which is known only by pain in the region of the uterus, however slightly occurring, the delivery will terminate in immediate death.

Therefore, if the bold, hasty, rash, careless, impatient, impetuous, imprudent and unthinking practitioner*, should forcibly attempt to deliver, before the moving fibres of the uterus are capable of contracting, as soon as the child &c. is thus hastily extracted, the uterus will be left as an empty flaccid bag, its blood vessels will remain open, and a violent profuse hæmorrhage or flooding will immediately commence, and

---

* Such practitioners I have known.

the woman in confequence will almoft inftantly faint, and expire under his hands *. *Woe to fuch practitioners! Horror to their reflecting minds!*—Whereas, on the contrary, no fatal misfortune to the woman would enfue, nor difquieting reflections invade the practitioner's mind, if he had but allowed patience and attention to have been his guides, and efpecially in thofe cafes of uterine hæmorrhages happening in the later months of uterogeftation, and arifing particularly from a fmall portion or lobe of the placenta lying clofe to, or over the os uteri; for when that orifice begins to expand and dilate, as towards the laft month of uterogeftation it always doth, then, in confequence of that dilatation, a feparation of certain veffels of the placenta from the uterus begins to take place, as is manifeft firft by a flight fhew of blood, which after a little time ceafes, then on any flight motion of the woman again appears; then ceafes, and again returns, or perhaps continues thus trifling for feveral days; then

* And fuch accidents have happened under their hands.

again

again appears, so as to alarm the woman and her friends.

The practitioner must now be upon his guard, he must sooth the woman by speaking in a mild gentle manner to her, by telling her, that all will soon be well, but that much depends on keeping herself at rest, quiet in mind, &c. and slight restringents, as ~~tinctura~~ *Infusum* rosæ, or even a little cold water very slightly acidulated with vinegar, may now and then be given. To her friends he may hold out, that there may be some danger, nay much danger, yet he hopes all will terminate well; but let him ever remember, that much depends on his own conduct; he must not be impatient; he must watch the woman very attentively and carefully, perhaps for some days, even many days*, for a contraction of the moving

---

* I have had several such hæmorrhages from such a cause continue more or less for two or three weeks, nay even longer, and yet do well; labor pains at last coming on, the hæmorrhage has stopt, and the child expelled naturally; and when the placenta has been expelled, I could then see how great a portion had been separated by its

containing

ing fibres of the uterus taking place, which he may know by flight and alternate pains coming on the woman, alfo by her uneafy motions of body, by the tenfenefs, in the time of pain, of the membranes, which may perhaps now be felt through the opening of the os uteri; if fo, he may now be well affured, that all may do well; but ftill he muft give time, for if he does not, he may as yet lofe the woman; for though the difcharge is perhaps fometimes profufe, neverthelefs, as the pains increafe in force, thofe motions or contractions of the uterus may fo clofe up the bleeding veffels as in fome degree to fupprefs the flooding; and when the os uteri is fufficiently dilated, he may break the membranes; or if the placenta lies entirely over the os uteri, and thus prefents itfelf, he may pierce through it, in order not merely to difcharge the waters, but to give more eafe to the contracting fibres of the uterus; and as the waters

containing much coagulated blood, and by its being of a more dark livid colour than the other parts of the placenta.

difcharge,

discharge, the tonic state and contraction of the uterus will more and more take place, and thus very much suppress the hæmorrhage or flooding; but observe that when the practitioner has broken the membranes, or pierced thro' the placenta, as the labor pains, or the natural alternate motions and contractions of the uterus, will undoubtedly very soon or immediately increase in their force and bearing down as the waters are discharged; if then the head presents naturally, the rest of the labor must be left to nature, and conducted as a natural labour*; but if the head

---

* In flooding cases, at the latter end of uterogestation, at the very onset of, or during labor, it is generally advised not to wait for the natural efforts of the pains, but immediately on breaking the membranes, &c. to deliver by the feet, whatever may be the presentation; but I have conducted very safely many cases of this kind, where the head presented, entirely by trusting to the natural labor pains; and to this I have been induced, first, to avoid certain accidents that may occur in turning a child and delivering it hastily by the feet:—but, secondly and chiefly, from considering the state of the circulation between the mother and child; from which it appears, that the discharge of blood in such uterine hæmorrhages proceeds more from the vessels of the detached portion of the placenta, than from the denuded vessels of the uterus;

for

head should be stopt in the pelvis by any particular distortion of it, or by the largeness of the child's head, or by any other unforeseen

for though when the placenta lies on part of, or over the os uteri, as that orifice dilates, it separates a certain part of the cervix uteri from a certain portion of the placenta; nevertheless the blood cannot proceed so freely from the vessels of the cervix uteri as from the placenta, although granting, as the os uteri dilates, that the vessels of the cervix uteri are not contracted; yet it must be allowed, that they are then so far extenuated as to be rendered in their diameters much smaller than they were before, and that some are drawn so fine in their diameters as even not to admit of red blood to pass through them, consequently in this their effect, no great degree of flooding can proceed from them, but on the contrary, from the vessels of the separated portion of the placenta the blood may proceed freely, but yet not so much, nor so readily from the venal part of it as from the arterial, though granting that the veins of the detached portion of the placenta anastomose with those veins of the placenta not yet separated from the uterus; yet as such detached veins, immediately on their separation from the uterus, must lose their power and become inactive, and as the blood to occasion an hæmorrhage must pass through them in a retrograde course, and contrary to their shape and figure, &c. it consequently will soon stagnate and coagulate therein, so as to stop them up, and thus no great discharge can continue long to pass through them so as to be deemed a flooding: but it is from the arterial

vessels

forefeen caufe, then the child muft be delivered as in retarded labors. But if the head fhould not prefent when the membranes,

veffels of the feparated part of the placenta, that the greateft fhare of the hæmorrhage proceeds, and confifts of that blood which would have been returned from the child to the mother, if there had not been a feparation of a certain portion of the placenta from the uterus; for as the blood is taken up from the uterus into the placenta by the venal ramifications of the vein of the funis, and along which it paffes and proceeds into the body of the child, and then almoft immediately paffes on and circulates through the heart, &c. it is from thence almoft as immediately returned through the aorta and iliacs into the arteries of the funis, and thence into their branches in the body or fubftance of the placenta, and in this courfe the blood can pafs readily, fo that from the detached portion of the placenta the feparated arteries may pour out their blood fo very profufely, as to occafion fuch a difcharge as to be deemed a flooding. Hence though the lofs of blood in this manner may in fome time equally affect the mother, as if it had proceeded entirely from the denuded veffels of the uterus, yet it is not fo very immediately quick as to require an immediate delivery at its firft appearance, when the os uteri near to or at the onfet of labor firft begins to dilate; and hence as fuch difcharges of blood from the uterine parts do not appear to me to proceed fo much or fo very immediately from the mother as from the child, nor to be fo very immediately dangerous to the mother as fome have imagined, fo as to require

branes, &c. are broken, and some other part is felt, the child must be delivered by the feet, as in preternatural cases, taking care,

quire an immediate and forcible delivery, I have, therefore, from these considerations, trusted such labors attended with a discharge of blood owing to a partial separation of the placenta, and particularly if the head presented, as much as possible I could to nature, and have as often not been disappointed in my expectations. But granting the flow of blood to be chiefly discharged from the arteries of the detached portion of the placenta, it may again be said that immediate delivery is also absolutely necessary on the child's account, as that may soon perish from the immediate loss of blood; but not comparing the loss of the child's life to the importance of the mother's life, it is evident that this will not be the case, as the child is constantly supported with blood from the mother so as to compensate for the loss of blood that is discharged by the detached arteries of the placenta, except when more than the major part of the placenta is separated, and even then the child will not so immediately expire as to require for its preservation an immediate and forcible delivery; for when the major part of the placenta is separated from the uterus, the hæmorrhage in proportion thereunto rather ceases, and besides the os uteri is then so much dilated, and the head descended so far down in the pelvis, as to shew that the labor is so far advanced, that the expulsion of the child by the natural pains of the woman will be effected before its expiration can take place; and again, as it must be allowed that the child has, even though the placenta is totally separated

care, whatever may be the mode of delivery in these cases, that it is performed slowly, gradually and regularly, with as little force

rated from the uterus, a vis vitæ, or a circulation of its own sufficient to support itself for some little time, so that provided there is no particular obstruction to its expulsion, or a preternatural presentation, a forcible delivery is not necessary; but if there is a peculiar obstruction, or a preternatural presentation, then artificial assistance is certainly very necessary, but not otherwise. Let those then, who advise immediate and forcible delivery in uterine hæmorrhage happening in the last months of uterogestation, at the onset of parturition, or during labor, consider well these reflections, and then I am persuaded that they will alter their opinion, and confess that immediate delivery at the onset of, or even during labor, is not so very immediately necessary as they have advised, and so very positively recommended; for there can be no show, or discharge of blood, or flooding, without a separation of some part of the placenta from the uterus; so in whatever part of the uterus the placenta is connected, whether on the os uteri, or near to the cervix, or towards or at the fundus uteri, still the same effect will take place as to the hæmorrhage, which will in every case proceed more immediately from the placenta and child than from the uterine vessels of the mother; and this will appear very evident to any practitioner who will observe in those cases of retarded or preternatural labors where there are previous to its birth evident signs of the child being dead, as when from the funis umbilicalis falling through the os uteri down before the presenting

part

force as possible, and with the greatest cautions.

I am part of the child into the cavity of the vagina so as to be touched, and even sometimes to protrude so far through the os externum as to be seen, and then from not perceiving any pulsation in its arteries, from its feeling cold, and from its appearing of a greenish brown colour, it is evident that there is a total cessation of the circulation between the mother and child, and that the child is dead, and further, that it is verging to a state of putrefaction, and, in consequence thereof, that there is an entire separation of the placenta from the uterus; nevertheless under all these circumstances no profuse hæmorrhage or flooding is found to take place, although the child is still within the uterus, and keeping its sides expanded; for it is to be observed, that when by any morbid cause either from the mother or from within itself the child loses its life before its birth, that then, as its circulation has ceased, no hæmorrhage or flooding can ensue through the child from the vessels of the placenta, neither can it from the vessels of the uterus, for as the putrefaction of the child and placenta comes on, so the placenta in consequence thereof must separate from its attachment to the uterus, the placenta now being an offensive dead substance, the uterine vessels will withdraw and contract themselves therefrom, and entirely cast it off, and that without any ensuing hæmorrhage or flooding, the vessels of the uterus now being in a contractile state, as is very evident in such cases from the placenta so immediately following the expulsion of the child, and more particularly

I am now, fourthly,

To obferve with refpect to convulfions happening in labor, that they are the moft alarming

larly in thofe cafes where the fœtus included and wrapt up within the membranes unruptured is entirely and wholly expelled altogether from the uterus, as often is feen in mifcarriages even in the later months of uterogeftation; this certainly is a pofitive proof that in partial feparations of the placenta from the uterus the hæmorrhage or flooding proceeds more immediately through the child from the detached veffels of the placenta than from the uterine veffels of the mother, confequently is not fo very immediately dangerous to the mother as to require fo very immediate and forcible a delivery as has been fo generally fuppofed, and fo very injudicioufly recommended in order to fave the woman's life, but which I am certain on the contrary has deftroyed numbers, for whenever there is any hæmorrhage or flooding proceeding from the uterine veffels, the uterine fyftem muft be in an atonic ftate, ought then forcible delivery to take place in fuch a ftate? Certainly no; though it is fuppofed by practitioners that fuch a proceeding will fo irritate the uterus as to recover its tonic ftate, and thus contract the bleeding veffels, and ftop the hæmorrhage; but this is a very dangerous, uncertain and random practice, and I believe inftant death to many. Hence may readily be perceived the reafon why reftringents are in general fo inefficacious in fuch uterine hæmorrhages; perhaps ftopping up the vagina with foft rags wetted with cold water will prove the beft and moft effectual reftringent;

alarming accident that can poffibly occur; as they evidently fhow that the nervous fyftem is in a very high degree of excitement, and that the habit or difpofition of the woman is in an exceffive ftate of mobility and irritability.

As convulfions generally come on very unexpectedly even in the moft natural labor, the practitioner ought to be very circumfpect, confiderate and attentive; and firft he fhould endeavour to afcertain whether the convulfions are truly epileptic or hyfteric, as for their removal fome difference in their treatment is neceffary; for as the latter are not by far fo dangerous as the former out of which very few recover, fo our prognoftics to the woman's friends muft be regulated.

To diftinguifh of what kind the convulfions are, I fhould be inclined to fuppofe, even though there was a continued lofs of fenfe, that they were of the hyfteric kind,

reftringent; nay the only certain one that can in fuch cafes be ufed until the expulfive pains of labor come on, and thefe forcibly taking place expel the fœtus &c. And thus happily the hæmorrhage ceafe.

by the frequent working of the eye-lids, by the almoſt conſtant motion of the jaws, by the full globular ſwelling and tightneſs of the throat threatening ſtrangulation, by the frequent alternate riſings of the thorax, by the deep ſighings or rather ſobbings, the toſſing and beating of the arms, and particularly if ſlight remiſſions and renewals of thoſe ſymptoms occur; but if there are ſmall intervals of ſenſe, or remiſſions of the moſt violent ſymptoms, the convulſions may then be deemed truly hyſterical, and as they frequently ceaſe before, or generally ſoon after the labor is over, the woman may perhaps ſafely recover.

But on the contrary, I would rather ſuppoſe that the convulſions were of the epileptic kind, if the eye-lids rather remained ſhut, and continued motionleſs, with the eyes turned up ſo far under the upper lids as almoſt to hide the cornea, if the muſcles of the face are ſo affected as to diſtort and draw the face more to one ſide than the other, with frothy matter foaming from the mouth, the jaws as it were fixed, the tongue ſwelled and ſeemingly too large for the

the mouth, great difficulty in breathing, stertor, insensibility, &c. and lastly, from not perceiving any kind of remission of any one symptom; hence much danger may justly be apprehended and predicted, few recovering, and indeed in some cases it is greatly to be suspected that apoplexy terminates the whole affair.

Whatever may be the cause of convulsions at the onset of labor, whether owing to so great a mobility of habit, that, when the first motion in the uterine system is given for labour, it is communicated through consent to every muscular fibre of the body, so as to excite or throw it into convulsions;—or whether, from the gradual dilatation or expansion of the uterus during gestation, there is at the period of labor a loss of tone in the uterine system from excessive distension, so that to remove this atony and to recover a due tone and vigor in the moving fibres of the uterus, nature rouses her utmost efforts, and in the general commotion convulsions take place; or whether convulsions, happening at the onset of or during labor, are owing to any other cause, I cannot say.

But I believe, that, for the most part, their approach may in some measure be foreseen, and their attack suspected, and this particularly if the woman complains of pain, or frequent giddiness and swimmings in her head, and especially if the pain increases so as often to be violently lancinating, or if she complains of dimness of sight, or of seeing very indistinctly, with a frequent wavering mist coming before her eyes, or if she has sudden and violent rigors following her labor pains, with a flushed bloated face, often turning a little blackish; and also if she has a kind of spasmodic stricture across the thorax occasioning a very great shortness and difficulty in breathing; thus some one or more of these symptoms may proceed, increasing more and more violently, until suddenly there comes on a total abolition of sense, with strong convulsive motions in every part of the body.

With respect to the management requisite in such dreadful situations, as the convulsions must be considered as of the acute kind,

kind, dependent upon exceffive irritability and mobility of the nervous fyftem, the treatment muft chiefly turn upon quieting the nervous fyftem, and obviating the effects of its convulfive exertions; hence the means to be adapted for thefe purpofes fhould be—If the convulfions are of the hyfteric kind, firft to give the woman a free admiffion of cool air, next to empty her bowels with fome moderate purgative clyfter, and then try to allay the convulfive emotions with antifpafmodics, as tinct. caftor. ʒi. fpt. æther. vitriol. comp. g^{tt}. xxx. tinct. opii g^{tt}. xv. in a little aqua pura, and which may be repeated occafionally until the hyfteria ceafes, which commonly happens before the labour is over; and I have known this medicine of great fervice in labor, where a kind of hyfteric ftricture coming acrofs the thorax has occafioned a very oppreffive breathing, greatly diftreffing to the woman, particularly in time of labor pain.—But if the convulfions are of the epileptic kind, we muft not only admit of cool air, but even fprinkle her frequently with cold water; and as there is more or

lefs

less of an inflammatory diathesis and plethoric habit merely from contraction of vessels, some blood, in proportion thereunto, should immediately be taken away.—Leeches to the temples, if we were certain that the bleeding after their falling off would be profuse, would have very good effect; and particularly so, if there were present any approaching symptoms to apoplexy, as sometimes is the case.—Cupping, with scarifications on the head, would also be very serviceable; but as the evacuation by these means is generally slow and insufficient, we should always have recourse to bleeding from the arm or jugular vein, by which means we may not only prevent a rupture of the vessels in the head, but also remove a dangerous suffocation of the lungs; and further, to assist these means, the bowels should be emptied by a brisk purgative saline glyster; after which, to remove the general inflammatory spasmodic affection of the epileptic convulsions, the sedative antiphlogistics with antispasmodics may be given, as vin. antimon. tartar. g^{tt}. xx.—lx. tinct. opii g^{tt}. xx. with spt. lavend. comp.

comp. g^{tt}. x. in a little aqua pura, and the doses should be repeated as the urgency of the symptoms seem to require, until some nauseating rather than an emetic effect of the medicine takes place, or the convulsions cease; but, for the most part, all our assistance proves ineffectual, for generally the woman remains insensible, and dies just before or soon after her delivery.

Such then is the treatment from medicine in convulsions occurring at the onset of labor; but with respect to the delivery in such cases, much is to be left to nature: the practitioner must be very careful how he excites pain by any unnecessary assistance to promote delivery, lest he thereby increases the convulsive exertions: he may indeed at the invasion of the convulsions touch the woman, but with a view only to know the state of the os uteri, how rigid, soft, or how far disposed to dilate, but no particular assistance* is requisite until the

os

---

* Immediate delivery is generally recommended in convulsions at the onset of or during labor, but I doubt much of its propriety; at the onset of labor it cannot be performed

os uteri is much dilated, and the child's head sunk far down into the pelvis; and even then, if the child is not retarded, and the pains sufficiently strong, it must still be left to be expelled by the efforts of nature, as in a natural labor; but, when the uterine parts are sufficiently dilated, if the head should be retarded by any means, it may then very slowly and gradually be extracted with the Forceps, or, if the child should present preternaturally, then the operator must very gradually, slowly, and with very little force, introduce his hand into the vagina; and on breaking the membranes, &c. he must proceed on, and with all due cau-

---

performed without much force and injury, as the os uteri is not then sufficiently dilated, and when labor is somewhat advanced, if the convulsions are of the hysteric kind, it is certainly unnecessary, as the convulsive emotions commonly subside before the labor is over; and in convulsions of the epileptic kind, as there is generally such a strong spasm throughout the muscular fibres of the uterine system, it will be difficult even to effect a delivery without much force, &c. (hence increase of irritation and consequently increase of spasm and convulsion) until the os uteri is entirely dilated by nature; and then, if the head presents, the child will soon be protruded into the world without any assistance.

tion, &c. deliver the child by the feet, as in other preternatural cafes; after which the feparation of the placenta muft be left to nature, and then finally extracted as in a natural labor.

If now the woman fhould continue to live, the greateft care muft be taken of her, total filence muft be obferved round her, no attendant or vifitor fhould unneceffarily be admitted, and the room fhould be kept dark, and of a moderate temperature (viz. between 50° and 60°) her food fhould confift of thin foft gruel, of which only a few fpoonfuls fhould be given at a time, and mild dofes of fedative antifpafmodics, as camphor, caftor, mufk, &c. with the fpt. æther. vitriol. comp. &c. in a little aqua pura, may be given her until the convulfive affection is entirely ceafed and fenfe returned, or death has clofed the fcene. How far blifters may be of fervice in fuch cafes, I cannot fay; they are recommended, and I have feen them applied even on the head, but without any good effect, and with the fame inefficacy I have known the warm bath ufed.

Thus

Thus much for convulsions during labor.

It must now, fifthly,

Be observed with respect to the common or usual modes of managing natural labors, as desiring the woman to bear down and force with her pains; giving her warm stimulating liquors to increase the frequency and strength of them, and likewise for the practitioner to pretend to assist, as it is called, with attempts to dilate the uterine parts, and also when the child is passing through the os externum to extricate it quickly and forcibly, and after the child is born, to attempt as precipitately, immediately, and with some force, to extract the placenta, are, I say, modes, though too frequently used by impatient and unthinking practitioners, excessively wrong, absurd, and productive of the greatest mischief, nay, of every misfortune that can happen to a lying-in woman; whereas if the labor was entirely left to nature, nay rather retarded, all would terminate well.

Hence let me say, that a woman, at the onset

onset of her labor, should be left entirely to herself, to move where she pleases, but should keep herself cool; she should be spoke to in a mild soothing, though cheerful manner, and she should be particularly desired not to use any effort in forcing her pains in order to accelerate her labor, for her pains are now the natural and necessary pains dilating the os uteri, &c. and were she to use any forcing or bearing down with them, the uterine parts would not so readily dilate, but, on the contrary, the natural course of her labour would be much perverted and prolonged; all that the woman has now to do, is to keep herself extremely quiet, to take very little food only occasionally, when thirsty a spoonful or two of thin soft gruel, lest it should excite sickness*; and she should commit herself entirely

---

* As the involuntary actions of vomiting agitate and affect every part of the body, so more particularly during labor such emotions always produce sudden and involuntary forcings of the uterus, and excite the woman to bear down violently, greatly distressing, painful, and often injurious to herself; therefore, if sickness should accompany

tirely to the moderate and natural efforts of nature, and not to be anxious about the return of pain, or the tediousness of her labor; for in due time the uterine parts will expand and dilate, so that the child, without her forcing or bearing down, or any other way striving or straining herself, will gradually, with very little pain, ~~more ease,~~ and with the utmost safety to herself, be delivered.

Hence it is evident, that the practitioner at the onset of labor should attempt nothing; and so far from being solicitous for the delivery, that he should avoid touching the woman; and he should wait patiently

pany labor, it should be suppressed, if possible, by giving her a little magnesia alba or creta prepared, with a few drops of tinct. opii, or spt. æther. vitriol comp. in a spoonful of aqua pura, and she herself should be cautious in her eating and drinking, so as not to load or offend the stomach; a few spoonfuls occasionally at a time of any thin soft gruel, will be sufficient to support her through labor, and thus she will avoid much fatigue, and the consequences of violent and involuntary strainings, forcing and bearing down of the uterus, and every injury that can happen from such emotions of her stomach and body. And hence I will say, contrary to the adage of old women, that a sick labor is not a safe nor the best labor.

until

until by the tone of her voice, by the motion of her body, and by the frequency and continuance of the returning pains, it is evident, that her delivery is drawing near; he then may touch her in order to know the ſtate of the uterine parts, how open they are expanded, how far the child is advanced, and how ſituated. If then every thing is found to be natural, even though the labor may be far advanced, it muſt be left to go on by the natural efforts of the labor pains; for the practitioner muſt be very cautious in touching, leſt he breaks the membranes too ſoon*; and he muſt not even pretend to lend any aſſiſtance, or to uſe any force, or attempt to dilate the parts, for the greateſt ſecret in midwifery is to know when artificial aſſiſtance, however gentle, is not neceſſary, but will do hurt; all then, that the practitioner has now to do, is only to wait patiently, not to touch the woman, but to exhort her to be as quiet as poſſible, and not even to bear

---

* That is, not before the os uteri is dilated as large as, or almoſt as large as the child's head, and which ſhould be far deſcended into the pelvis.

down, strain or force herself, or even to use any particular efforts with her natural labor pains with a view to forward her delivery; for so far from being of service to her will such unnecessary exertions of her pains be, that they will rather pervert her delivery, and occasion it to be much more excruciating and distressing to herself; nay, it will prolong it, as such exertions will greatly fatigue her mind and body, and will render the next returning pains irregular, so that the natural dilatation of the uterine parts will not gradually go on, nor properly take place; for nature knows best how and when to unravel and unfold the fibrous structure of the os uteri, vagina, perinæum, &c. so as to dispose them for such a proper dilatation and expansion as to give an exit for the child; therefore if labor is hurried in the dilatation &c. of these parts, numbers of fibres must be greatly injured, and perhaps ruptured, so that they cannot recover their proper places after the delivery; hence, from such loss of tone, &c. many diseases and complaints to females arise, as *procidentia vel* prolapsus uteri, vaginæ, &c. consequently every

every labor, that is, where the head prefents naturally, fhould be left as much as poffible to nature, even though it fhould continue lingering &c. for feveral days, and then no doubt but all will terminate well, and the woman foon recover.

Whereas, on the contrary, if the practitioner is impatient, and injudicioufly defires the woman to affift her pains by holding her breath, at the fame time by forcibly ftriving and bearing down; and befides this, if he very officioufly attempts to affift with his hand in touching her, and thereby endeavours, however flowly, to dilate the uterine parts, he will greatly hurry the woman, and he will not only inflame the uterine parts, but will fo irritate them as through confent of parts perhaps to throw the woman into hyfterics, but at leaft to excite ficknefs and vomitings; and thus occafion the woman to force and bear down violently and involuntarily, fubverting the natural and fpontaneous dilatation of the os uteri, vagina, perinæum, &c. fo very proper for the woman's fafe delivery and recovery.

And now, granting, that the os uteri is sufficiently dilated, that the membranes are broken, and the waters are discharging, and that the head is so far advanced as to begin to bear upon the anus, nay upon the perinæum, &c. nevertheless the practitioner even now, instead of encouraging the woman to bear down, should particularly desire her to the contrary, and to use no more efforts to forward her delivery than the mere natural bearings of her pains; nay he should desire her rather to suppress them; for at this period of labor much attention should be paid to the woman, for the slower or less forcing the pains are, the perinæum, &c. will have more time given it as gradually to distend; and the os externum will then so expand and gradually enlarge, that the child's head, instead of being forced, will rather gently and slowly slip through it into the world: the practitioner, therefore, at all times in natural labors should never touch the woman with any other view than to know how labor advances; his assistance in any other manner is unnecessary; for, if he now attempts to

dilate

dilate the perinæum, &c. he will do much mifchief. Hence as the head protrudes through the os externum preffing on the perinæum, and greatly extending it outwards and backwards, the practitioner fhould only apply the palm of his hand on it, in order to give it fupport; which muft be regulated according to the woman's bearings, &c. and he fhould now very particularly exhort the woman even not to bear with her pains; for if fhe does, and particularly with a fudden and forcible exertion, he fhould then as forcibly prefs his hand againft the perinæum, for without fo doing and retarding the fudden expulfion of the child's head, it is ten to one, but that the perinæum by fuch fudden bearing exertions will burft, and be rent even into the anus, *a dreadful confequence!* Whereas by the woman's retarding, or rather as much as poffible fhe can fuppreffing or withholding her pains in their bearing down, the child's head will flowly flide through the os externum without hurting the perinæum, &c. in any degree whatever, or any accident whatever occurring.

The child's head being thus flowly and fafely delivered through the os externum, the further deliverance of the child's body muft not be immediately attempted, nature muft be waited on. She has now more to do internally within the uterine fyftem, and particularly within the cavity of the uterus, than what fome practitioners may think; for by the flow and gradual delivery of the head through the os externum, the neceffary contraction of the fundus uteri as gradually takes place, thereby the placenta as gradually begins to be feparated from the fundus uteri, and that without any enfuing flooding; and further, by letting the body of the child reft, or rather retaining it within the vagina during the next natural returning pain or few pains of the woman, the fundus uteri will further contract, which gives a more fafe, perfect and natural feparation of the placenta: all this a good, humane and careful practitioner fhould confider within his mind, fo that when the natural pains of the woman come on for the expulfion of the child's body, he muft not attempt officioufly to affift the delivery

livery thereof by pulling at the head to extract the shoulders, &c. but, on the contrary, he should rather prevent the sudden and forcible expulsion of the shoulders, by supporting or rather moderately holding the child's body within the vagina against the violent expulsive efforts or forcings of the labor pains; but when the pains recede, then to let the child's body, as of its own accord, slowly and gradually pass through the os externum into the world; hence in a natural labor the woman should be delivered merely by her own natural efforts or pains, and which she herself should not attempt to increase by any exertion, as holding her breath, forcing or bearing down, nay by rather restraining or withholding her pains, her uterine parts will gradually enlarge, distend and open, and every part of the child will pass or rather glide through the os externum, &c. slowly and gradually, with very little pain to herself, and without lacerating the perinæum, or any hurt or injury whatever happening; and besides these advantages, as there will be time for the fundus uteri, &c.

&c. gradually to contract, and thereby flowly, perfectly and naturally to feparate and caft off the placenta, no undue retention of it will happen, or any violent flooding enfue therefrom.

Whereas, on the contrary, if the woman is violent in her pains, bearing or forcing downwards with all her ftrength; or more particularly, if the practitioner is impatient to deliver the body of the child immediately after the head is born; and particularly, if it happens to be a fmall child, and the uterus much diftended with water, he will prevent the natural, gradual and fpontaneous feparation and expulfion of the placenta; for generally the uterus will contract where there is leaft refiftance to it; fo when the child is very fuddenly delivered, the contraction will always be towards the cervix uteri (for as yet there is an adhefion between the fundus uteri and the placenta), and occafion what is called the hour-glafs contraction of the uterus, retaining the placenta as within a purfe, with its mouth drawn clofe together by its ſtrings: and now, if the practitioner is not very

very circumspect and careful, he will create much mischief, and even occasion death itself; for if he is in a hurry to extract the placenta, not giving time for a proper separation of it, and relaxation of the hourglass contraction of the uterus, or not thinking that there may be a twin child* behind, if he pretends very unthinkingly to extract the placenta by pulling at the funis with some degree of force, if that is weak and slender it will for the most part break, and leave the body of the placenta behind; and then perhaps, if he is impetuous, he will be so very imprudent as rashly to introduce his hand into the uterus; and if then he finds not another child within the

---

* As a woman may have twin children, it is necessary soon after one child is born, the funis umbilicalis tied and cut, and the child delivered to the nurse, for the practitioner to touch the woman, in order to know, before he thinks on the extraction of the placenta, if there be a twin child remaining within the uterus; if so, and the head presents, the delivery of it must be left to nature, and to the natural efforts of her pains; but if any other part presents, it then must be delivered as in preternatural labors; after which the practitioner must leave the separation of the placenta to nature, and then to be extracted as in other natural labors.

uterus, he will violently tear off, and separate the placenta, and thus extract it, leaving the connecting or anastomosing vessels of the uterus with the placenta ruptured, open and bleeding; hence the woman will soon become faint from loss of blood, and hence will arise many other very dangerous and deplorable diseases often terminating in death. Again, if there is no twin child within the uterus, and the funis is so strong as to withstand breaking on being forcibly and violently pulled, then there will ensue either a very disagreeable bearing down of the vagina, or, as I have seen, a total and complete inversion of the uterus, and from which death has soon followed.

Therefore, I would advise every practitioner to proceed slowly and tenderly in delivery, to give time for a gradual, natural and necessary contraction of the fundus uteri, so that the perfect and natural separation of the placenta may take place; and further, for the safe and total expulsion of the placenta, the practitioner should, after the child is born (although there is no twin child behind in the uterus), wait patiently for some time, that it may be separated by nature, and protruded from the uterus into the vagina by the natural pains of the woman; and indeed by a slow and gradual conducting of the labor, the placenta generally will be safely, gradually and perfectly separated from the fundus uteri, and will then for the most part be naturally protruded so far as to fall down into the vagina almost immediately after, or as soon as the child is delivered; and if so separating in time of labor, and laying in the vagina, it then seldom causes or creates much after pain, nay little or no pain; therefore, after the practitioner has waited for some time after delivery and no returning after-pains of the woman occur, as generally in such cases do not (all which observation and experience will soon teach him to know), he may then touch the woman; and on finding the placenta laying within the vagina, he may then be assured, that the fundus uteri is properly contracted, and that the placenta is perfectly, totally and naturally separated therefrom, and is descended into the vagina, and that by a gentle bearing

down

down of the woman, as if going for stool, and at the same time the practitioner gently drawing the funis downwards, the placenta may be very safely extracted, and slipped through the os externum, and thus the woman will be safely delivered.

But if the placenta does not separate from its adhesion to the uterus almost immediately after the child is delivered, and cannot be felt in the vagina by the practitioner's finger when introduced into the vagina, he must then wait patiently until the separation of the placenta takes place by a still further contraction of the uterus, which is known by pains alternly coming on, which are the more close contractions or natural closing up of the uterus; if then, after having waited for ten, fifteen, twenty, nay thirty minutes, for a few after-pains to contract the fundus uteri, and to cast off the placenta, so that it may be sunk down into the vagina and there readily touched with the finger, it may then be safely assisted along at the next occurring pains, after the manner just mentioned, and thus extracted. But if no pains occur after the child is de-
delivered,

delivered, and the placenta cannot as yet be touched with the finger, it muſt be left to nature, and the practitioner muſt wait even if it is for ſeveral days, two, three, four, or even longer, avoiding by every means the raſh and rude introduction of the hand into the uterus *, for ſeparating it, or I may ſay tearing or rather flaying it off with the finger from the tender and irritable ſurface of the uterus. Therefore if the gentleſt efforts to extract the placenta do not ſucceed, and prove ſufficient, it muſt be left entirely to nature; for after a few days it will probably be expelled naturally, and thrown out through the os externum, and this I have known greatly to be aſſiſted by the gentle efforts or expulſive bearings as at ſtool, occaſioned merely by the return even of an emollient clyſter.

Such then are the modes † by which labors

---

* Though ſo generally, but yet ſo injudiciouſly, adviſed in adheſions, or rather retentions of the placenta within the uterus.

† With reſpect to the eſpecial manner how each operation in the different kinds of labor ſhould be performed

labors fhould be conducted, and fuch are the accidents, injuries, &c. that may happen to women injudicioufly treated and hurried in their deliveries. Thus much for the management of natural labors and fafe extraction of the placenta.

It will now, fixthly,
Be proper to mention, that there is yet another accident which may happen to a lying-in woman, and which fome practitioners may not think of, nay perhaps never imagine, and which is equally, if not more

formed particularly, the author has not here attempted to explain, that will be the fubject of future differtations, comprehending alfo a general review of the feveral opinions of the different authors of midwifery, as to the particular conduct and management of pregnant, parturient and lying-in women; but in this, as a kind of introductory or general differtation thereunto, he has only endeavoured chiefly to fhew and point out the neceffity, whatever may be the mode or inftrument neceffary for the delivery, that it fhould be conducted flowly, gently, gradually and regularly, with patience and humanity, avoiding every degree of unneceffary force, leaving the whole entirely, or as much as poffible, to the efforts of nature.

dangerous

dangerous than any other, perhaps from not being suspected; and that is from certain parts of the uterus through the hurry and violence of labor irregularly and unduly contracting, by which means certain portions of the uterine vessels are as it were encircled and constricted as in a ligature, hence from such constricted portions of the uterus arise many uterine complaints, as excruciating after-pains, inflammation, mortification with all its dreadful effects; and hence the source perhaps of every child-bed fever so very frequently fatal. How slow, how regular, how gradual, attentive and cautious, ought then every practitioner to conduct even the most natural labor.

Hence, to prevent any painful, severe, or dangerous complaint happening to women during or after their parturition (and particularly from too great hurry or officious assistance to hasten delivery), every kind of labor should be conducted slowly, gradually and gently; and the woman should be even constantly exhorted not to bear down, or force with her pains, or use any other means to increase the natural efforts

forts of her labor; so that the child may be delivered gently, gradually and slowly, and with as little force as possible; and the separation, &c. of the placenta should in like manner be left to nature, it being ten to one but that some mischief will arise, if the practitioner attempts to hurry, or imprudently assumes according to the labor to assist beyond what is necessary; for it has ever been observed, that rash, impetuous and injudicious practitioners in midwifery have always the greatest number of bad cases, and that arising merely from their own too hasty, rash, imprudent and bad management; and indeed how should it be otherwise, as they never consider nor reflect on the cases that they attend, and perhaps are totally ignorant how to manage them? It is coolness in temper, constant assiduity, attention and length of time only, that can make a good, safe and experienced practitioner in midwifery..

In fine, what I have mentioned in these few pages I can aver to be true, from a practice of more than thirty years, and from the attendance on more than five thousand

thousand * labors; out of these, I could relate many instances or cases to illustrate what I have here asserted, but omit them as being tedious, and perhaps similar to what other practitioners in midwifery may have also seen, probably without paying any kind of attention to them; but if any one doubt what I have here asserted, let them for the future observe, and remark every case that they may attend, and then I am confident they will in a little time find,

* This number perhaps to some persons may appear great, and too many to fall to the practice of one man; but the author must inform them that for upwards of twenty years there was no other male practitioner in midwifery than himself in Oxford, and but very few round it for many miles, consequently many very tedious, dangerous, laborious and preternatural cases fell to his attendance and care, all which he endeavoured to conduct after the manner here related, and indeed such modes and treatment, as relates to the mother, child, and placenta, he has for many years inculcated amongst his medical friends; and if it had not been for the delay of waiting for the occurrence of a number of flooding cases at the onset of or during labor to have satisfied his mind from whence the hæmorrhage really proceeded, and to ascertain positively the best and most safe mode, &c. of conducting such cases, a somewhat similar dissertation to this, as relative to its other parts, would have appeared some years ago.

that what I have here related is nothing but the truth.

Thus much for the conduct and management of labors in general; but as the Obstetric Forceps is of such consequence in the delivery of retarded labors, it will now,

Seventhly,

Be neceſſary to attempt geometrically to proportion and delineate the different forms of the ſingle curved Obſtetric Forceps, as mentioned in the foregoing pages; and firſt with reſpect to the common ſingle curved Forceps as delineated in

Fig. 1.

Which ſhews the two ſides of the ſingle curved Forceps connected or locked together, conſequently can only exhibit the thickneſs of the blades, their proper curvature, the breadth of their ſhanks, together with the length, thickneſs, ſhape, &c. of the handles.

But to form the ſingle curved Obſtetric Forceps geometrically, and as near as poſſible

sible to the alterations of Dr. Smellie's Forceps, as mentioned page 10, and thereby to fix it to one general shape, size or standard. First, as in Fig. 1, draw in length a right line as A B eleven inches and one half of an inch, which will be the axis of the Forceps; then from the point A, set off six inches and one half of an inch towards B as at the point C, which line A C divide in its middle as at the point D, with the line E F, at right angles with the line A C; then on the line D E, set off one inch and the four tenths of an inch as at the point G, and again on the line D F, set off one inch and the four tenths of an inch as at the point H; then with the radius of four inches and one half of an inch on the line G F as from G to J, draw a curve line from A through G to C, and again, with the like radius on the line H E as from H to K, draw another curve line from A through H to the point C, and thus two regular curves as A G C, and A H C (the segments of a quadrant), for the internal curvature of the Forceps' blades are formed.

And now, to give the blades their proper expansion at their ends, from the point A set off on each curve line A G, and A H, the six tenths of an inch as at the points L and M, which are the ends of the Forceps' blades, and thus the ends of the blades gain an expansion of about the eight tenths of an inch as from L to M, and hence the length of the chord line of each blade when locked together will be, from the point C to the points L and M, about six inches and half the tenth of an inch.

And further, to give the blades their proper thickness, through the points L and M, draw a right line as the line N O, and then from the point L toward N set off two tenths of an inch as at P, and again from the point M towards O set off two tenths of an inch as at Q; then from the point G on the line G K set off two tenths and half a tenth of an inch as at the point R, and then on the line R J set off from R four inches and four tenths of an inch as from R to S, and on that radius draw a curve line from R to the point P; again, from the point H on the line H S, set off two tenths and half the tenth of an inch as at the point
T,

T, and then on the line T K set off four inches and four tenths of an inch as from T to U, and on that radius draw a curve line from T to Q, and thus the thickness of the blades from their utmost curvature, as at the points R G and H T, is regularly formed to their ends at P L and M Q, which from their outer edge must be gradually rounded off to their inner as at L and M.

Again, to give the full length and curvature of each blade, first set one foot of the compasses at the point J, and with the other foot continue the curve line from G through C to the six tenths of an inch as to the point V; and again, set one foot of the compasses at the point K, and with the other continue the curve line from H through C to the six tenths of an inch as to the point W;—then to give the shanks of the blades their proper breadth from their utmost curvature as at R G and H T, with a radius of six inches two tenths and half the tenth of an inch on the line R F, as from R to the point X, draw a curve line from R to the point W as the curve line

R W; then with the like radius of six inches two tenths and half a tenth of an inch on the line T E as from T to the point Y, draw a curve line from T to the point V as the curve line T V, and thus the major part of the shanks of the Forceps' blades are formed.

But to give the shanks their final length and breadth, set off from the point C on the line C B one inch as at the point Z; then from the point W draw a right line to the point Z as the line W Z; and again, from V to Z draw another right line as the line V Z, which two right lines terminate the length and breadth of the shanks of the Forceps' blades, and also from the sloping part of the groove for the locking part of each blade, as from W to Z, and from V to Z; and thus, are geometrically ascertained, the length and proper curvature of the Forceps' blades, together with their thickness from their ends to their utmost curvature, and from thence the breadth or width of their shanks as necessary for their strength in their different parts even to their locking parts, and also the slopes of of the grooves for the locking parts of the Forceps.

As to the formation of the Forceps' handles, I can only say positively that in length from the point C to their ends as at B, they should not exceed five inches; or from the point Z (the bottom part of the groove for the locking of the Forceps' blades) to B four inches, so that the whole length of the Forceps shall not exceed eleven inches and half the tenth of an inch: as to the width of the groove for the locking part, it should be only just so wide as to admit easily the thickness of the shank of the opposite blade; and the cheek *a* of the locking part should be of such a thickness as one tenth of an inch, or what is necessary only to give them strength, so as to keep the shanks of the blades steady (vid. Fig. 2): but these particulars will always vary a little to the fancy of the maker, as will also the size and shape of the handles, for as the handles of the Forceps if they were made entirely of solid iron would be very heavy, therefore the maker, as soon as he has formed the cheek *a*, and the breadth of the groove for the locking part and its thickneſs (which may be two or three tenths of an inch as from the points W and V to the points $b, b,$) of ſufficient ſtrength, ſize, &c. he then beats the remaining part of the iron into a thin plate as at $c, c, c, c,$ ſo as not to be more than one tenth, or hardly to be more than half the tenth of an inch thick at the bottom or end of the handles as at B: the other part is made up of hard wood or horn, rivetted to it as at d, d, d, d; and then it is cut, ſloped and ſhaped, to the fancy of the maker, who for the moſt part finally covers it with leather.

But to fix in ſome reſpect the ſize, ſhape and thickneſs of the handles; firſt, through the point B at right angles with the line B Z C D and A the axis of the Forceps, draw the right line ef, and then on that line ſet off one inch on each ſide of the point B, as at the points g and h; then draw a right line from the point g to the point W, the top of the groove for the locking part of the handle, and the like on the other ſide from the point h to the point V,
and

and thus is obtained a regular flope for forming the Forceps' handles; and further, to regulate the indentations in the handles neceffary for the operator's firm holding them, fet off from the point g on the line g W, firft half an inch, then another half inch, and then one inch more, and another inch beyond that towards W, as at the points i, k, l, m; and the like on the other fide from the point h on the line h V, as at the points n, o, p, q; then let the maker in forming the thicknefs of the handles gradually round and indent them from the points W and V, to the points m and q, about one tenth of an inch, and from whence let him again gradually fwell them out to their full extent as at the points l and p, from thence again gradually round and indent them to the points k and o, two tenths of an inch, and from thence again gradually let them fwell out to their full extent as at the points i and n, and then gradually round them off to their ends, as reprefented by the black wavering line from the points $b, b,$ to the bottom of the handles at B.

Hence

Hence from this description and view of Fig. 1, I believe that any artift may form geometrically the proper curvature of the Forceps' blades, their thickneſs, and the length and width of the ſhanks in all their ſeveral parts as neceſſary for their ſtrength, together with the regular ſlopes of the grooves for the locking parts, and the ſize, ſhape and thickneſs of their handles.

Fig. 2.

But, as it is impoſſible to ſhew the neceſſary breadth of the Forceps' blade, the thickneſs of its ſhank, and the width of the groove for the locking part, and height and thickneſs &c. of the cheek of the locking part, together with the proper curvature of the blade, &c. in one figure, it is therefore neceſſary to repreſent them in another as in fig. 2, and even here as it is impoſſible on paper to repreſent any depreſſion or curvature in a foreſhortened manner, that is, by a right line, ſo the breadth of the blades, &c. can only be given as correſpondent to the different parts of their curvature, as in Fig. 1, and therefore thereunto to conſtruct geometrically

geometrically the breadth of the Forceps' blade, the thickneſs of its ſhank, with the width of the groove for the locking part, the height and thickneſs of its cheek, and breadth of the handle, firſt draw a right line eleven inches and half the tenth of an inch in length as the line A B in Fig. 2, which is the axis of the Forceps; then from the point A on the line A B ſet off the ſeven tenths of an inch as at the point C, and then on that point, and with that radius deſcribe the ſemicircle D A E, after which draw its diameter D E, and this forms the ſhape and breadth of the blade at its end; then ſet off from the point C on the line C B two inches and one tenth of an inch as at the point F, and croſs it at right angles with the line G H, and from the point F on the line G F ſet off half an inch and the half tenth of an inch, and the ſame diſtance on the line F H as at the points J and K, and thus the breadth correſponding to the utmoſt curvature of the blade is formed: then from the point F on the line F B ſet off two inches and three tenths of an inch as at the point L, which is the point of the

blade's

blade's bifurcation, or divifion of the furrounding iron which forms the blade; then from the point L on the line L B fet off the fix tenths of an inch as at the point M, through which point at right angles draw the line N O, and then on each fide of the point M on the line N O fet off one tenth and half the tenth of an inch as at the points P Q; then draw a right line from the point P to the point J, and likewife from the point J to the point D; and again, draw a right line from the point Q to K, and from K to the point E, and thus the breadth of the Forceps' blade is formed from its end to the narroweft part of its fhank. But as this blade, if it were compofed of one folid iron plate, would be very heavy, it has therefore, to make it lighter and ftill to have the fame effect, been cuftomary for the maker to take out the inner part, leaving only a furrounding border about the breadth of two tenths or three tenths of an inch: but, however, to form this part of the Forceps' blade, firft fet off from the point J on the line J F two tenths and half a tenth of an inch; and the fame diftance from the

point

point D on the line D C; and the same from A on the line A C; and then from E on the line E C; and again from K on the line K F, as at the points R, S, T, U and V; then draw a right line from the point L to the point R, and from R to S, and then describe the semicircle S T V, on the radius of S C; and then from the point U draw a right line to the point V, and from V to L, and thus the proper breadth of the surrounding part or iron of the Forceps' blade is ascertained.

And further, to form the thickness of the bottom part of the shank of the Forceps' blade, &c. first set off from the point M on the line M B half an inch as at the point W, from which point to the left hand draw at a right angle with the line W B a right line as the line W X, and then set off half an inch from the point W on the line W X as at the point Y, and again set off one tenth of an inch more towards X as at the point Z, this gives the height and thickness of the cheek of the locking part;—then from the point W on the line W B set off a quarter of an inch as at the point *a*, from which

draw

draw to the left hand at a right angle with the line aB the line ab, and on which from the point a set off one tenth and half the tenth of an inch as at the point c, and then from c set off three tenths and half the tenth of an inch more towards b as at the point d, which gives the width and top of the groove for the locking part; then from the point a on the line aB set off the six tenths of an inch as at the point e, through which point at right angles draw the line fg, then on each side of the point e on the line fg set off one tenth and half the tenth of an inch as at the points h and i, then set off three tenths and half the tenth of an inch from the point h on the line hf as at the point k, which gives the width of the groove at the bottom of the locking part; and then from the point k on the line kf set off one tenth of an inch as at the point l, which fixes the thickness of the cheek of the locking part. Hence, to form out the cheek of the groove for the locking part, draw a right line from the point l to the point Z, and from Z to Y, and from Y through d to k; and to form the width of the

the groove at the top of the locking part, draw a line from *d* to *c*, which is the top of the groove; and to form the width of the groove at the bottom of the locking part, draw a line from the point *k* to *h*, which is the bottom of the groove; and finally, to form the thickneſs of the ſhank of the Forceps' blade, draw a right line from the point *h* through *c* to the point P; and then from the point Q draw a line to the point *i*, and thus far is formed the breadth of the Forceps' blade, the thickneſs of its ſhanks, the width of the groove for the locking part, and thickneſs of its cheek.

But with reſpect to the further conſtruction of the Forceps' handles, it can only again be obſerved, that the length of the handle ſhould be four inches from the point *e*, or bottom of the groove for the locking part, to the end of the handle at the point B; and in its width towards its bottom it may be about one inch and three tenths of an inch. However to form the breadth of the handle ſo as ſomewhat geometrically to anſwer to its thickneſs in Fig. 1, draw through the point B at right angles with the

the line *e* B the line *m n*; and from the point B on the line *m* B set off nine tenths of an inch as at the point *o*; and again on the line B *n* set off four tenths and half the tenth of an inch as at the point *p*; then draw a right line from the point *o* to the point *l*, and again from *i* to *p*; and thus is obtained a regular slope (for forming the handle) from the bottom of the groove for the locking part to the end of the handle; and further, to make it correspond with the shape of the thickness of the handles in Fig. 1, set off on the line *o l*, first, half an inch, then another half inch, and then one inch, and then another inch more towards *l* as at the points *q, r, s* and *t*; and the like distances on the other side as on the line *p i* as at the points *u, v, w* and *x*, on that line; and further, to regulate the necessary indentations of the handle to answer agreeably to those in Fig. 1, let the maker in shaping the breadth of the handle gradually round and slope it on each side from the points *l i* to the points *t x*, about half the tenth of an inch; then from the points *t x* to the points *s w* as gradually swell it out to the full extent; then again, gradually

gradually round and indent it on both sides about one tenth or one tenth and half the tenth of an inch to the points $r\,v$; and then again gradually swell it out to the full extent at the points $q\,u$; and lastly, as gradually round it off to the end, as represented by the wavering black lines from the points l and i to the bottom of the handle as at B.

And further, with respect to the other parts of the Forceps, the maker must ever remember carefully to smooth, polish and round off every part, so as not to leave any kind of sharp, prominent, or acute edge whatever; and thus by observing and minutely following these directions and geometrical proportions, one general structure of the Obstetric single curved Forceps may by any artist be easily constructed and made; which in the hands even of the most inexpert practitioner, will rarely hurt or bruise the child's head, or compress it so in its delivery as to occasion its death; and never in the hands of the expert and judicious practitioner will any way injure either the mother or child.

Fig. 3.

But as there are some cases, particularly those mentioned from page 15, 16 and 19 to 23, that require, for the safe delivery of the woman and child, one blade of the Forceps to be much more narrow than those of the single curved Forceps already described and delineated in the two foregoing figures, it will be here necessary to illustrate such a blade, as in Fig. 3; first it must be observed, that this narrow blade must be made of an entire plate of iron, and must have the same thickness, the same curvature, the same sized shank, groove, cheek, &c. for the locking part and handle as the single curved Forceps delineated in Fig. 1 and 2; but the breadth of the blade must not be more than an inch wide towards its top, and about the eight tenths of an inch at its utmost curvature, from which to the narrow part of its shank it must gradually decrease in its breadth to the three tenths of an inch.

But, to construct it geometrically, first draw a right line eleven inches and half the tenth of an inch, as the line A B in Fig. 3, which

which is the axis of the Forceps; then from the point A towards B set off half an inch as at the point C, and on that radius describe the semicircle D A E, and then draw its diameter as the line D E; after which, from the point C on the line C B set off two inches and three tenths of an inch as at the point F (which corresponds with the utmost curvature of the blade as in Fig. 1), and immediately cross it at right angles with the line G H, and then on both sides of the point F set off on that line four tenths of an inch as at the points J and K; again, from the point F on the line F B set off two inches and nine tenths of an inch as at the point L, and at right angles cross it with the line M N, and then on each side of L set off on the line M N one tenth and half the tenth of an inch as at the points O P; after which, from the point O draw a right line to J, and from the point J to D, and then again from P to K, and from K to E, and thus the breadth of this narrow blade of the single curved Forceps will be constructed; and as the construction of the other parts are the same as described and delineated in Fig. 1 and 2, to complete the whole of this side of the Forceps, the artist is referred thereunto, and when finished will readily join either with the side of the Forceps Fig. 2, or with the reflected blade Fig. 11, 12, whenever such cases as are mentioned from page 15, 16 and 19 to 23 occur, so as to require its application.

Fig. 4.

But as there may also happen other such cases as are mentioned from page 15, 16 and 19 to 23, that will require a much narrower blade of the Forceps than even that of Fig. 3, for its ready introduction into the pelvis, and safe delivery of the woman and child; therefore to form such a blade it was necessary to divide one side of the single curved Forceps into two parts, and which are joined together at the point of the blade's usual bifurcation by the contrivance of a little hinge, and by diverging the handles a little therefrom the two fangs will open and shut at pleasure, so that after the fangs are introduced into the pelvis in their shut or close state, they will afterwards again open

open and spread on the child's head, and have equally as much effect as the entire blade of Fig 2, or of any other kind of Forceps.

But to delineate this divided or fanged side of the single curved Forceps geometrically, first draw a right line eleven inches and half the tenth of an inch as the line A B in Fig. 4, which is the axis of the Forceps, then set off from the point A five inches and one tenth of an inch, as at the point C (which is the center of the hinge which joins the two parts together), and then on the radius of A C, describe through the point A the segment of a circle as D E, and then from the point A on the segment D A set off seven tenths of an inch, and the like distance on the segment A E as at the points F and G, which is the utmost expansion of the fangs, and nearly equal to the breadth at the top of the blade Fig. 2; and further, from the point F on the segment F A set off three tenths of an inch, and the like from the point G on the segment G A as at the points H and J, which is the breadth of each fang at its end.

Then to delineate the hinge of the two fangs, firſt, through the point C, at right angles with the line A B, draw the line K L, and from the point C on each ſide of it on the line K L ſet off one tenth and three tenths of a tenth of an inch as at the points M and N ; after which, from the point C on the line C A ſet off two tenths and the two tenths of a tenth of an inch, and the like from C on the line C B as at the points O and P ; then on the point M ſet one foot of the compaſſes, and on the radius of M N, from the point O through N to P, deſcribe the ſegment of a circle as the ſegment O N P ; and again, ſet one foot of the compaſſes on the point N, and on the radius of N M deſcribe from the point P through M to O the ſegment of a circle as P M O, and thus the ſize, ſhape, &c. of the hinge, which connects the two parts of this ſide of the Forceps together, is formed *. Then from the point C on the line

---

* To illuſtrate more perfectly the nature of this hinge, and to ſhew the manner how the two parts of this ſide of the

line C B set off the six tenths of an inch as at the point Q, and cross it at right angles with the line R S, and from the point Q

on

the Forceps are thereby particularly connected together; first, let it be understood that Fig. 5 and 6 exactly represent the size of the hinge, and part of the fangs and shank of Fig. 4, unconnected, by which it is obvious that on each part over the center C there is a semielliptical knob to be raised, as O N P in Fig. 5, and O M P in Fig. 6, each correspondent with Fig. 4, and out of which the several parts of the hinge are to be formed.—Secondly, it must be understood, that through the middle of the knob of Fig. 5, sideways and longways in direction from the fang part to the handle there must be cut a notch two fourths of its breadth to the center as from O through C to P, and that then of the same breadth there must be further indented a semielliptical groove similar to the semielliptical shape of the knob, and so as to make the whole cavity represent the shape of a myrtle leaf, as O M P N in Fig. 4.—And thirdly, it must also be understood, that the upper and under side of the knob over the center C of Fig. 6, must be also cut away sideways and longways in the direction from the fang to the handle, each one fourth of its thickness to its center C, as from O through C to P, and that then they are to be further indented so as to receive the semielliptical cheeks of the cavity of Fig. 5, and thus the middle part of the knob of Fig. 6. will appear as a knuckle in shape to a myrtle leaf, which knuckle of Fig. 6 is to go into the

on each side of it on the line R S set off one tenth and half the tenth of an inch, as at the point T and U; and then from the point T to the point F draw a right line, and the same from the point C to the point H, and again from C to the point J, and once more from the point U to the point G, and thus the breadth of the two fangs expanded is formed with the thickness of

cavity of Fig. 5, and when they are so fitted as to move easily on each other, then through the center C they are to be rivetted together. But, further to shew the nature of the hinge connecting the two parts of this side of the Forceps together, let it again be understood, that Fig. 7 and 8 represent exactly the breadth of the shank and thickness of the knob where the hinge is formed, so that when the cavity of the groove of Fig. 5, and the knuckle of Fig. 6, are formed, that then the cross section of the cavity of Fig. 5 will appear very similar to the notch in Fig. 7, and that the cross section of the knuckle of Fig. 6 will appear very similar to the projection of Fig. 8, which projection *a* of Fig. 8 being put into the cavity *b* of Fig. 7, and rivetted together through their centers C, the joint of the hinge will appear as in Fig. 9; hence may easily be perceived the construction of the hinge connecting the two parts or fangs of this side of the Forceps together.

some

some part of the two shanks of this side of the forceps.

But further to illustrate the other parts of the shanks, and the formation of their handles, together with the groove of the locking part, its cheek, &c. belonging to the two parts of this side of the Forceps, it will be proper to refer the artist to

Fig. 10.

Which shews the two fangs shut close together, and the handles expanded. But to delineate them geometrically—first, draw a right line, which is the axis of the Forceps, eleven inches and half the tenth of an inch as the line AB Fig. 10; then from the point A on the line AB set off five inches and one tenth of an inch as at the point C, the center of the hinge joining the two parts together, and on the radius of C A describe through A the segment of a circle as DAE, and then on each side of the point A, on that segment, set off towards D and E three tenths of an inch as at the points F and G, which is the breadth of each fang at its end, making the whole breadth at the ends when

thus

thus closed together six tenths of an inch; then at the point C describe, as in Fig. 4, the hinge on which the two fangs move and are connected together. Again, from the point C on the line CB set off the six tenths of an inch as at the point H, and through that point at right angles with the line CB, draw the right line JK, and on that line on each side of the point H set off two tenths of an inch as at the points L and M, then from the point L to the point F draw a right line, and another from the point M to the point G, and thus is formed the breadth of each fang, whose end at their outward points F and G must be a little rounded off as in Fig. 10. And further, to form the remaining parts of the shanks and the handles of this side of the Forceps when expanded, first, from the center of the hinge at C set off on the line CB five inches nine tenths and half the tenth of an inch as at the point B, and then through B describe the segment of a circle as NBO, and then on each side of the point B set off five tenths of an inch as at the points P and Q; then draw a right line from the point P to C,

and

and from Q to C, which lines are the inside lines of the two handles to this side of the Forceps, and which gives between them the proper distance for the expansion of the handles when the fangs are shut, and which gives an equal expansion to the two fangs when the handles are shut as in Fig. 4.

But further, to form out the shank, handle, &c. to the fang of the left part, first set off from the point C on the line C P one inch nine tenths and half the tenth of an inch as at the point R, from which at a right angle to the left draw the right line S R, and then from the point R towards S, set off one tenth and half a tenth of an inch as at the point T, then draw from T to the point L a right line as the line T L, and thus it forms out and finally finishes on this part the thickness of the shank down to the bottom part of the groove for the locking part; but as the groove and cheek of the locking part and the rest part of the handle on the left part of this side of the Forceps are the same as those of Fig. 1 and 2, the artist is referred thereunto in finishing this part of the Forceps.

ceps. But finally, to form out the shank and handle to the fang on the right part of this side of the Forceps, from the point C on the line C Q set off one inch nine tenths and half the tenth of an inch as at the point U, and from which at a right angle with the line C Q, draw to the right hand the line U V, and then from U towards V set off one tenth and the half of a tenth of an inch as at the point W, then draw a right line from the point M to the point W, and this finally forms the thickness of the shank of the right fang as far down as to be even with the bottom of the groove for the locking part of the left fang; after which, the artist may beat out the iron to its proper length and thickness, so as to be at the bottom of this handle not above half the tenth of an inch thick as at X, X, and to which a piece of hard wood or horn as Y, Y, must be rivetted, and then finally formed, shaped, &c. according to Fig. 1 and 2.

Fig. 11.

With respect to Fig. 11, it shews the reflection of one blade of the Forceps on the other

other blade as mentioned page 34. Its application is to supersede in many cases of retarded labors, if possible, the use of those very dangerous and horrible instruments the blunted hook, crotchet, and such like.

But to delineate it geometrically proceed as in Fig. 1. And when its several parts are marked out, then as in Fig. 11, to reflect one blade, first, set one foot of the compasses on the point C, and from the point V on the radius of C V, the six tenths of an inch, draw a curve line from V until it bisects the curve line C H as at the point a; then from the point G on the line G H set off two tenths and half the tenth of an inch as at the point b, and then on the radius of b J draw a curve line from the point b towards the point L (the full end of the single curved blade of Fig. 1.), so as to be correspondent with the line G L, to about three fifths of the curve line GL, or one inch and eight tenths of an inch from G towards L, as at the points *c* and *c*, which fixes the end of the reflected blade; and thus the thickness of the upper part of the reflected blade is also given as from the points G b to the points c c, the end of which must be

rounded

rounded off as in Fig. 11; and further, from the point b set off three inches and three tenths of an inch on the line b J as at the point d, and then draw a curve line from the point b to the point a as the curve line b a, which forms the breadth of the reflected shank of this side of the Forceps; but to give room for the reflected blade to join readily, easily, and unite evenly with the other blade, from the point C on the line C a, set off one tenth and half a tenth of an inch as at the point e, and then on the line G J set off four inches and one tenth of an inch as at the point f, and on that radius describe a curve line from G to the point e, as the curve line G e, and thus a little space as at g will be left between the two shanks, so that the shank of the reflected blade will readily move over the other as that of Fig. 3. And to make their blades lay evenly together, the maker must give their shanks a little bend sideways over the groove, and indeed he must ever remember, that there is to be such a little bend (which cannot be described by line, or hardly be perceived by sight) always given to the shank of the

other

other blades, as is neceſſary to ſet the ends of the blades directly oppoſite to each other; and laſtly, to finiſh this reflected blade of the Forceps, the curve line G C muſt be continued through the point V one tenth and half the tenth of an inch as at the point h; and then from the middle of the curve line V_a as at the point i draw a right line from i to h, and thus a little hook is formed, whoſe point h muſt be a little rounded off, and thus it ſerves as a catch for the ſupport of the reflected blade on the upper part of the groove for the locking part belonging to the oppoſite ſide of the Forceps, by which means the reflected blade is kept ſteady, and the handles of the Forceps are ſo firmly locked together, that they will not ſeparate nor ſlip from each other in extracting the infant's head; for it muſt be obſerved, that the action of the reflected blade of the Forceps is not on its concave ſide, that ſide has no power merely of itſelf on the head of the infant as being within it; and indeed it has no power on the head except as ſerving only by its convex ſide to preſs the bones and integuments of the ſkull cloſe together
againſt

against the ~~convex~~ *concave* side of the narrow blade Fig. 3, that is applied over the outside of the head, so that a firm hold is supported and kept between them on the integuments, &c. of the head, which could not ~~be~~ so regularly, or to so good a purpose, *&* accomplished by the operator's hand merely by squeezing or pressing the handles together, if there was not this catch to support the convex side of the reflected blade against the ~~convexity~~ *concavity* of the other blade placed over the outer part of the child's head: hence, by this catch preventing the reflected blade from slipping down when placed within the skull, and by thus enabling the blades to keep between them a firm hold on the integuments and bones of the skull, and likewise assisting the ~~convexity~~ *concavity* of the blade Fig. 3, placed on the outer part of the head, to act as a kind of hook or vectis thereon, the head may probably be drawn down and extracted without having recourse to either of those terrible instruments the blunted hook, crotchet, or any such like.

FIG.

Fig. 12.

As to Fig. 12, it shews the reflection of the blade separately and distinctly by itself. But to delineate it geometrically without referring to Fig. 1, first, draw a right line eleven inches and one half of an inch as the line A B in Fig. 12, and from the point A set off six inches and one half of an inch as at the point C; then divide the line A C in its middle as at D with the right line E F at right angles with the line A C; then on the line D E set off from D towards E one inch and the four tenths of an inch as at the point G, and again on the line D F set off from D towards F one inch and four tenths of an inch as at the point H; then from the point H on the line H E set off four inches and a half as at the point K, and on that radius draw a curve line from the point H to the point C as the curve line H C; and again, with the like radius on the line G F, as from the point G to the point J, draw a curve line from G to A as the curve line G A; and then on that line from the point G towards A set off one inch and eight tenths

tenths of an inch as at the point L, which is the end of the reflected blade; and then from the point G on the line G D set off two tenths and half a tenth of an inch as at the point M, and on the radius of M J describe a curve line towards A, so as to be correspondent with the line G L as the curve line M N, which gives the full thickness of the upper part of the reflected blade as from G M to L N, the end of which is to be rounded off as in Fig. 11; then from the point C on the line C H set off one tenth and the half of a tenth of an inch as at the point O, and again from the point C on the line C H set off six tenths of an inch as at the point P; and then from the point M on the line M J set off three inches and three tenths of an inch as at the point Q, and on that radius describe a curve line from M to the point P as the curve line M P, and then from the point G on the line G J set off four inches and one tenth of an inch as at the point R, and on that radius describe from G to O the curve line G O, which forms the breadth of the reflected shank of this blade of the Forceps; and

further,

further, to form the reflection of the blade with its hook, fet one foot of the compaſſes on the point J, and on the radius of J C, defcribe from C downwards to the right hand a curve line feven tenths and half one tenth of an inch to the point S as the curve line C S, and then on the line C S fet off from C fix tenths of an inch as at the point T, and on that radius at C defcribe the curve line from T to P, which gives the reflection of the blade; and then to form its hook, from the middle of the curve line P T as from the point U draw a right line to the point S, and thus the hook (whofe point muft be a little rounded off) is formed for the fupport of the reflected blade when connected with that of Fig. 3. And finally, to finifh the fhank of this blade, fet one foot of the compaſſes on the point K, and continue the curve line from H to C through the point C fix tenths of an inch as to the point V, then from the point C on the line C B fet off one inch as at the point W, and then draw a right line from the point T to W, and from V to W, which is the end of the fhank of the reflected blade, and it alfo gives the proper flope for the groove of its locking part as from V to W, and thus is delineated geometrically the length, thicknefs and proper curvature of the reflected blade of the Forceps, with the width and reflection of its fhank, together with the flope of the groove for its locking part; as to its other parts, they are the fame as delineated in Fig. 1, to which for directions in their final conftruction the artift is particularly referred.

Fig. 13.

But further, with refpect to the reflected blade of the Forceps as delineated in Fig. 11 and 12, it is neceffary, for the fame reafons given for illuftrating Fig. 2, to mark out its breadth, &c. in another figure, as in Fig. 13. But firft it muft be obferved, that this blade muft be made of a plate of iron fimilar to that of Fig. 3 in every refpect except in length, *& reflection*, and as it is always to be ufed with that narrow blade, fo it is neceffary that they fhould correfpond in their breadth, though not in their length, for as the narrow blade Fig. 3 is always to be applied

plied over the head, and the reflected blade is always to be introduced within the skull, after it has been opened with a perforator, so the reflected blade is not required to be so long as the other by at least an inch, for reasons so obvious to every practitioner that it is unnecessary here to mention them.

But to construct geometrically the breadth of the reflected blade so as to correspond with the breadth of the narrow blade Fig. 3, first, draw a right line ten inches and half the tenth of an inch as the line A B in Fig. 13, which is the axis of the Forceps; then from A towards B set off four tenths and half a tenth of an inch as at the point C, and on that center and with that radius describe the semicircle D A E, and then draw its diameter D E; then from C on the line C B set off one inch, three tenths and half the tenth of an inch as at the point F, and cross it at right angles with the line G H; after which set off on that line on both sides of F four tenths of an inch as at the points J and K, then set off from F on the line F B two inches and nine tenths of an inch as at the point L, and draw a line through L at right angles with the line LB as the line MN; then on the line MN on both sides the point L set off one tenth and half a tenth of an inch as at the points OP; then from the point O draw a right line to the point J, and from J to D, and again from P to K, and from K to E, and thus the breadth of the reflected blade of the single curved Forceps is formed: as to the further construction of its other parts, they are the same as those of Fig. 2, to which the artist is referred for further instructions to finish this side of the Forceps, and when completed, will readily join with the narrow blade Fig. 3, whenever such cases occur as are mentioned page 34, so as to require its use.

Thus much for the geometrical construction, &c. of the different kinds of blades of the single curved Forceps as delineated in Fig. 1, 2, 3, 4, 10, 11, 12, and 13, from a view of which, and from a comparison of their several proportions with those of the real instruments, any practitioner in midwifery

wifery may readily perceive whether or not the maker has properly conſtructed them; but, however, in ſuch a compariſon, though it can be an error of no conſequence, yet it is neceſſary here to be obſerved, that if each figure of the Forceps was to be minutely and correctly examined by an abſolute meaſurement according to the deſcription given, there would perhaps in ſome of their parts be found a little difference in their geometrical delineation, as perhaps the half of a tenth of an inch, from what their proportions ſhould have been as mentioned in their ſeveral deſcriptions, and this ariſing from ſome little accidental incorrections of the engraver, but chiefly from the redrying of the paper after its having been neceſſarily damped for receiving properly the impreſſions of the figures from their engravings: but, if any artiſt will draw on a ſmooth even piece of paper any one or all of the figures of the Forceps as repreſented, according to the deſcription given, he will find, that the proportions are juſt, and that every part will correſpond, and that the figure or figures will then be perfectly and accurately delineated.

In fine, with respect to the figures, let it be observed, that Fig. 1 shews the real and absolute length, proper curvature, together with the proper thickness of the blades, the breadth of their shanks, the slopes of the grooves for the locking part, with the width of the cheek, and the size, shape and thickness of the handles.

Fig. 2, 3, 4, and 10, shew the different breadths of the different blades of the Forceps only as relative in length to the curvature of Fig. 1; but not their absolute length, as it is impossible to delineate a curve in a foreshortened manner by a right line, so the artist in making the different kinds of blades, as represented in Fig. 2, 3, 4, and 10, must only observe their different breadths as correspondent to the axis of the Forceps in respect to the length of their curvature, so in forming their breadth according to the length of their curvature as in Fig. 1, the artist must always measure from the shank beginning from the upper part

part of the groove of the locking part. As to the other parts of Fig. 2, 3, 4, and 10, they reprefent the thicknefs of the fhanks of the blade, and the breadth of the groove for the locking part, with the thicknefs of its cheek, and the breadth and length of the handle.

Fig. 5, 6, 7, 8, and 9, reprefent the mechanifm of the hinge connecting the two fangs of the divided blade of the Forceps.

Fig. 11 and 12 fhew the curvature, length, &c. of the reflected blade of the Forceps, the one as connected with that of Fig. 3, as Fig. 11; the other by itfelf as Fig. 12; and Fig. 13 fhews its breadth, &c.

But in fhewing the breadth of the different kinds of the Forceps' blades, it muft further be obferved, that by the pofition of the groove for the locking part, &c. to the left hand, the convexity of the curvature of the blade is fuppofed to be uppermoft, but in the reflected blade Fig. 13, it is fuppofed that the concave furface of the blade is uppermoft, therefore the artift in making the

the fide of the Forceps with its blade reflected muft remember this, or elfe he will perhaps fet the groove of the locking part with its cheek, &c. on the wrong fide, and confequently will revert the breadth of the handle, &c. and thus will give himfelf much trouble to alter it, as it will not then join with any of the other blades, and particularly with that of Fig. 3, along with which it fhould always be applied in fuch cafes as are mentioned from page 32 to 36.

How far I have aptly fixed the length, curvature, thicknefs, and other dimenfions of the feveral parts of the different forms of the fingle curved Obftetric Forceps to geometrical proportions, &c. I cannot fay, therefore I muft leave every practitioner in midwifery to judge of it as beft pleafes himfelf, conceiving in my own mind that geometrical proportions are the moft certain and beft proportions that can be obferved in the conftruction of any inftrument, machine, &c. as then one artift as well as another can always therefrom regularly and perfectly make it; and befides, it is evident, that when any machine or inftrument

ment can be thus conſtructed, it is more likely to anſwer in every reſpect the intentions of its conſtruction much better than if it had only been made, formed, ſhaped, &c. by any random method, as by ſight, cuſtom, or fancy, &c. of the maker.

Thus much for the geometrical proportions and delineations as neceſſary for conſtructing the different parts and various forms of the ſingle curved Forceps; it now remains only to be obſerved, that with five of the before-mentioned ſides of the ſingle curved Forceps, variouſly changed and connected together, five different pair of Forceps may be formed, each adapted to certain different and particular caſes of retarded labors.

As firſt, two blades of Fig. 2 connected together and forming the common ſingle curved Forceps as in Fig. 1, to be applied in ſuch caſes of retarded labors even in a well formed pelvis, where, from a wrong preſentation of the head, or from its largeneſs or length it reſts within the pelvis and cannot proceed any further.

Secondly,

Secondly, as the narrow blade Fig. 3 connected with the blade Fig. 2.

Thirdly, as the fanged blade Fig. 4 or 10 with the blade Fig. 2.

Fourthly, as the narrow blade Fig. 3 with the fanged blade Fig. 4 or 10.

Each of these particularly are to be applied according to the presentations of the child's head in such cases of retarded labors from certain peculiar distortions of the pelvis preventing the natural expulsion of the child's head.

And fifthly, as the reflected blade Fig. 12 and 13, connected with the narrow blade Fig. 3, as in Fig. 11, which is to be applied in such cases of retarded labors, arising, either from so great a distortion of the pelvis, or largeness of the child's head, as to require for the safe delivery of the woman the head to be opened, in which case the narrow blade Fig. 3 is to be applied over the head, and the reflected blade Fig. 12 is to be introduced within the skull, and then connecting or locking them together as in Fig. 11, a firm hold of the integuments and bones of the
skull

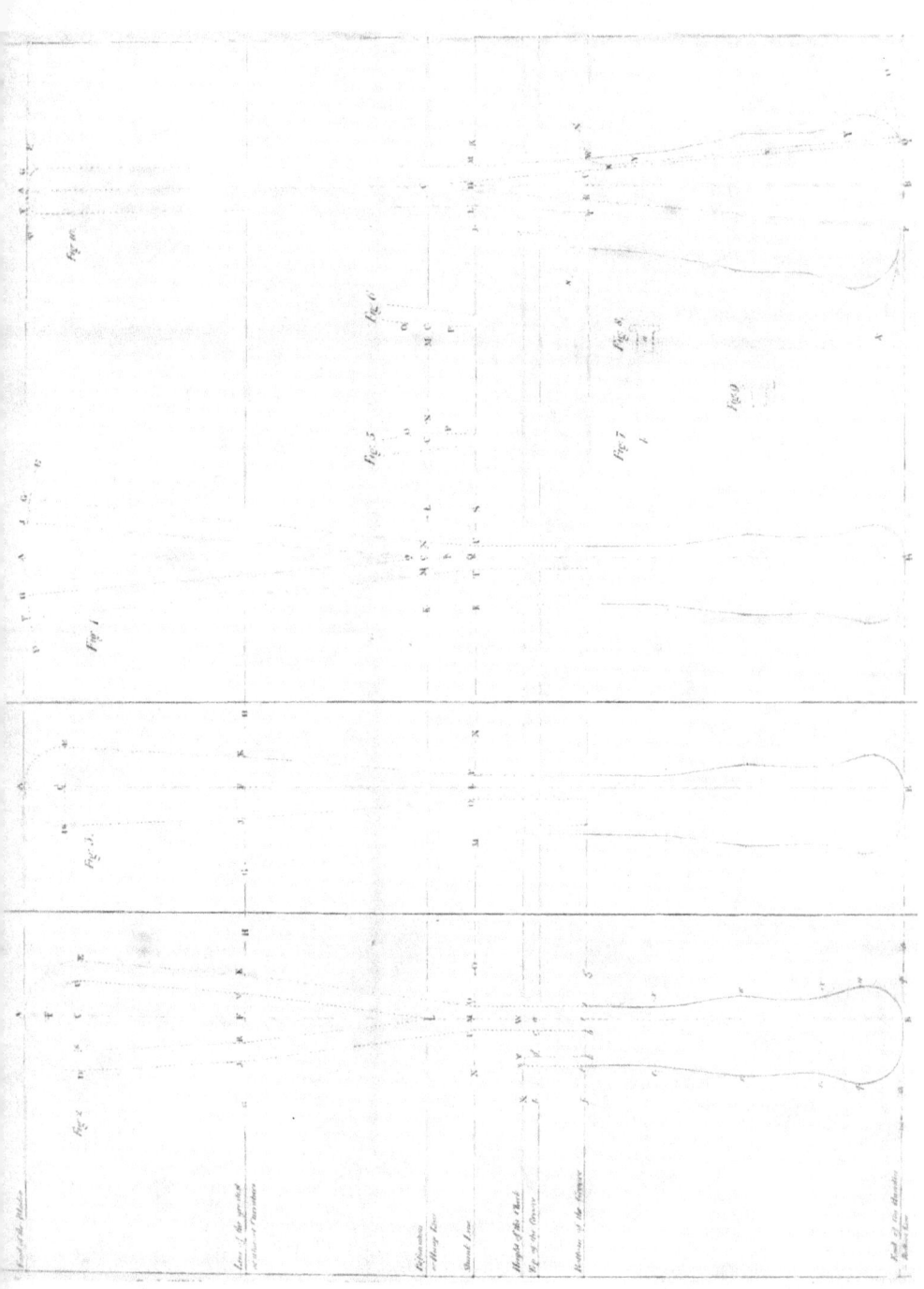

skull is secured, and thus the head may be drawn through the pelvis and delivered.

Such then are the several constructions, applications and various connections or junctions of the different sides of the single curved Forceps, which in every part I have endeavoured concisely to describe, explain, &c. and with accuracy to delineate geometrically.

And now, Sir, with all due respect, &c. let me subscribe myself,

  Your much obliged,

    humble Servant,

      R. RAWLINS.

THE END.

www.ingramcontent.com/pod-product-compliance
Lightning Source LLC
Chambersburg PA
CBHW020058170426
43199CB00009B/321